FORESTRY RESEARCH

A MANDATE FOR CHANGE

Committee on Forestry Research

Board on Biology
Commission on Life Sciences

Board on Agriculture

National Research Council

NATIONAL ACADEMY PRESS
Washington, D.C.   1990

National Academy Press • 2101 Constitution Avenue, N.W. • Washington, D.C. 20418

NOTICE: The project that is the subject of this report was approved by the Governing Board of the National Research Council, whose members are drawn from the councils of the National Academy of Sciences, the National Academy of Engineering, and the Institute of Medicine. The members of the committee responsible for the report were chosen for their special competences and with regard for appropriate balance.

This report has been reviewed by a group other than the authors according to procedures approved by a Report Review Committee consisting of members of the National Academy of Sciences, the National Academy of Engineering, and the Institute of Medicine.

This Board on Biology and Board on Agriculture report is based upon work supported by the U.S. Department of Agriculture [Forest Service and Cooperative State Research Service (Contract No. 88-G-034-1)], the Society of American Foresters, and the National Association of Professional Forestry Schools and Colleges.

Any opinions, findings, conclusions, or recommendations expressed in this publication are those of the authors and do not necessarily reflect the views of the U.S. Department of Agriculture, the Society of American Foresters, or the National Association of Professional Forestry Schools and Colleges.

Library of Congress cataloging-in-Publication Data

Forestry research : a mandate for change / Committee on Forestry
    Research, Commission on Life Sciences, Board on Biology [and]
    Board on Agriculture, National Research Council.
        p.   cm.
    Includes bibliographical references and index.
    ISBN 0-309-04248-8
    1. Forests and forestry—Research—United States. 2. Forest and
    forestry—United States. 3. Forests and forestry—Research.
    I. National Research Council (U.S.). Board on Biology. Committee
    on Forestry Research. II. National Research Council (U.S.). Board
    on Agriculture.
    SD356.5.F67   1990   90-6344
    634.9 '072073—dc20   CIP

Printed in the United States of America

Cover photograph: LANDSAT photograph of the Ocala National Forest in Florida, courtesy of Randy Kautz, Florida Game and Fish Commission.

COMMITTEE ON FORESTRY RESEARCH

JOHN C. GORDON (*Chairman*), Yale University, New Haven, Connecticut
WILLIAM A. ATKINSON, Oregon State University, Corvallis
ELLIS B. COWLING, North Carolina State University, Raleigh
MARY L. DURYEA, University of Florida, Gainesville
GEORGE F. DUTROW, Duke University, Durham, North Carolina
DONALD R. FIELD, University of Wisconsin, Madison
RICHARD F. FISHER, Utah State University, Logan
JERRY F. FRANKLIN, USDA Forest Service, University of Washington, Seattle
DAVID W. FRENCH, University of Minnesota, St. Paul
WILLIAM T. GLADSTONE, Weyerhaeuser Company (retired), Hot Springs, Arkansas
LAWRENCE D. HARRIS, University of Florida, Gainesville
LOIS K. MILLER, University of Georgia, Athens
JAMES R. SEDELL, USDA Forest Service, Corvallis, Oregon
RONALD R. SEDEROFF, North Carolina State University, Raleigh
DAVID B. THORUD, University of Washington, Seattle

NRC Staff

CLIFFORD J. GABRIEL, *Study Director*
CAITILIN GORDON, *Editor*
KATHY L. MARSHALL, *Senior Secretary/Project Assistant*

Advisor

JOHN ERICKSON, USDA Forest Service, Madison, Wisconsin

The National Academy of Sciences is a private, nonprofit, self-perpetuating society of distinguished scholars engaged in scientific and engineering research, dedicated to the furtherance of science and technology and to their use for the general welfare. Upon the authority of the charter granted to it by the Congress in 1863, the Academy has a mandate that requires it to advise the federal government on scientific and technical matters. Dr. Frank Press is president of the National Academy of Sciences.

The National Academy of Engineering was established in 1964, under the charter of the National Academy of Sciences, as a parallel organization of outstanding engineers. It is autonomous in its administration and in the selection of its members, sharing with the National Academy of Sciences the responsibility for advising the federal government. The National Academy of Engineering also sponsors engineering programs aimed at meeting national needs, encourages education and research, and recognizes the superior achievements of engineers. Dr. Robert M. White is president of the National Academy of Engineering.

The Institute of Medicine was established in 1970 by the National Academy of Sciences to secure the services of eminent members of appropriate professions in the examination of policy matters pertaining to the health of the public. The Institute acts under the responsibility given to the National Academy of Sciences by its congressional charter to be an adviser to the federal government and, upon its own initiative, to identify issues of medical care, research, and education. Dr. Samuel O. Thier is president of the Institute of Medicine.

The National Research Council was organized by the National Academy of Sciences in 1916 to associate the broad community of science and technology with the Academy's purposes of furthering knowledge and advising the federal government. Functioning in accordance with general policies determined by the Academy, the Council has become the principal operating agency of both the National Academy of Sciences and the National Academy of Engineering in providing services to the government, the public, and scientific and engineering communities. The Council is administered jointly by both Academies and the Institute of Medicine. Dr. Frank Press and Dr. Robert M. White are chairman and vice chairman, respectively, of the National Research Council.

Preface

The National Research Council (NRC) established the Committee on Forestry Research to create a vision of what such research must be like in the future in order for society to achieve desired forest management goals. As part of this vision, the committee addressed how forestry research benefits society in general, research needs of those who use forests, and the financial and human resources that must be called forth to meet those needs. The committee approached forestry research from an integrated, rather than a disciplinary, perspective.

The Committee on Forestry Research was sponsored by the U.S. Department of Agriculture's Forest Service and Cooperative State Research Service, the Society of American Foresters, and the National Association of Professional Forestry Schools and Colleges. We first met as a committee on January 30–31, 1989. Since that time we have solicited input from a wide variety of forestry and other sources, have held a workshop for interested organizations and individuals (Appendix A), and have reviewed other relevant NRC reports including *Managing Forest Genetic Resources* and *Opportunities in Biology*. We have attempted to examine all pertinent literature and have been given data by the Forest Service, Cooperative State Research Service, and others. This report is the carefully considered and discussed collective vision of the committee.

This report was developed in parallel with another NRC report—the Board on Agriculture's *Investing in Research*. The Committee on Forestry Research wrote *Forestry Research: A Mandate for Change* recognizing that the relationship between forestry and agriculture in this country (and else-where) needed to be enhanced and improved. We believe that the time has come to examine both fields and establish a new sense of partnership be-tween agriculture and forestry. This partnership must be based on a broad understanding of both fields and their essential similarities and differences. It will also require a new vision and renewed vigor in the research that supports both.

It is in the hope of a massive revitalization of forestry research that we offer this report as a catalyst for discussion and action.

JOHN C. GORDON, *Chairman*
Committee on Forestry Research

Contents

FORESTRY
RESEARCH

Executive Summary

Although concern about and interest in the global role and fate of forests are currently great, the existing level of knowledge about forests is inadequate to develop sound forest-management policies. Current knowledge and patterns of research will not result in sufficiently accurate predictions of the consequences of potentially harmful influences on forests, including forest-management practices that lack a sound basis in biological knowledge. This deficiency will reduce our ability to maintain or enhance forest productivity, recreation, and conservation as well as our ability to ameliorate or adapt to changes in the global environment.

To help overcome this unfortunate deficiency in knowledge, a new research paradigm will need to be adopted—an environmental paradigm. Even though previous approaches to forestry research employing the conservation and preservation paradigms have been adequate to meet many past forest management goals, they are now inadequate to guide forestry research into the future. However, the adoption of an environmental paradigm will require forestry research to increase the breadth of research areas covered and the depth to which they are investigated. Major issues that society faces concerning forests are

- how forests and climate affect each other, especially in the face of rapid global deforestation and forest degradation;
- loss of biological diversity;
- growing demand for wood, wood fiber, and derivative chemical products;

1

- increasing demand for the preservation of "pristine" forested areas;
- sustainable wood production integrated with protection of fish, wildlife, water, recreation, and aesthetic values; and
- maintenance of the health of forests nationally and globally.

All of these issues demonstrate the array of societal needs that depend on forestry research. From the protection of our vital forest products industry to the protection of regional and global environments, forestry research must be positioned to play a major role.

AREAS OF RESEARCH NEEDED TO ADDRESS MAJOR ISSUES

Five broad research areas critically need to be strengthened: (1) the biology of forest organisms; (2) ecosystem function and management; (3) human-forest interactions; (4) wood as a raw material; and (5) international trade, competition, and cooperation. These major research areas encompass the research needed to address current and future biological, climatic, and societal issues in forestry and the related management of renewable natural resources. Although the research needs described in this report have importance for tropical as well as temperate forestry, we do not divide research into those two categories, but highlight fundamental research areas that need attention globally. Basic forestry research should provide for

- understanding the basic biology and ecology of forests,
- developing information to sustain productivity of forests as well as protect their inherent biological diversity,
- designing and implementing landscape-level and other large-scale, long-term experiments,
- understanding the economic and policy-making processes that affect the fates of forests,
- developing systems of forest management that simultaneously produce commodities and maintain and improve environmental values,
- integrating the social component into research on forest ecosystems, which can then be applied to management practices,
- developing harvesting systems that recover timber values without degrading other values, and
- improving the efficiency of production and utilization of new and traditional wood products.

THE STATUS OF FORESTRY RESEARCH

Although much good research is now in progress within the forestry research community, in aggregate, forestry research is inadequate to meet society's needs. Forestry research must improve in quality and at the same time broaden its scope if societal issues are to be addressed adequately. In turn, the number of scientists participating in basic forestry research must increase. The number of scientists currently earning Ph.D.s in forestry and related fields has remained almost static since 1978, while the number of undergraduate degrees awarded has declined precipitously by approximately 50 percent. With increasing demands for forestry and forestry-related research, improvements are clearly required in the way scientists are recruited into forestry research. One way to enhance forestry's human resource is to promote interdisciplinary research, which will provide new technology and different research approaches. As forestry issues become more complex, the need for interdisciplinary research will become even greater.

With numerous advisory committees representing organizational research interests, leadership in forestry research has been fragmented. Government agencies and other organizations responsible for research activities can obtain policy advice from a wide variety of sources, such as internal advisory committees at the level of agency head or at the level of division. Research organizations can also draw upon other groups for advice, groups such as the National Research Council. Because of the broad range of research organizations and clientele of forestry research, none of the existing forestry advisory committees has adequately met the needs of the forestry research community in general. Therefore, a policy advisory mechanism must be established to provide leadership that transcends the interests of individual organizations.

Conclusions and Recommendations

• Establish a National Forestry Research Council (NFRC) to provide a forum for deliberations on forestry research and policy. The NFRC should be convened under the auspices of an organization or organizations that can facilitate discussion and action. Financial support for the council's activities should come from member organizations and other interested sponsors. The NFRC should consist of representatives from major organizations—such as government agencies, industry, conservation organizations, and academia—with strong interests in forests and related renewable natural resources and in agriculture. The NFRC would commission studies, conduct analyses, and provide advice to policy-makers on issues pertaining to forestry and related renewable natural resources.

• Encourage conservation groups and other nongovernmental organizations to more actively support basic forestry teaching and research through the activities of the NFRC.

• Provide a vastly expanded competitive funding mechanism to support active forestry scientists and to attract scientists not yet active in forestry research.

• Create centers of scientific emphasis integrating scientists supported by major forestry research organizations, including industry. This should be done for each of the five research areas. The creation of a center of emphasis does not necessarily require the construction of a new research facility. It does require, however, a mechanism for cooperative research that allows scientists to interact in a manner that enhances their productivity. More than one center could be established for each of the five research areas, depending on particular interests and strengths of the proposed center's participants. To create these centers, some existing research facilities may need to be closed and research funds may need to be redirected.

• Strengthen and broaden the teaching of forestry to attract a wider array of students, especially at the graduate level, and to interest other on-campus research groups in areas such as agriculture and biology.

• Increase the quality of forestry research by opening it to the broader scientific community.

• Establish research-management collaborations at large spatial scales. This would require interdisciplinary research on large tracts of land.

• Establish long-term forestry research (LTFR) grants to provide a peer-reviewed, competitive funding mechanism for long-term (greater than one forest rotation) research support.

• Establish competitive graduate fellowships for all areas of forest and environmental sciences.

• Develop a cadre of forest and related scientists that reflect the national and global population composition and that are equipped to solve domestic, international, and global problems.

SUPPORT FOR FORESTRY RESEARCH

The recommendations contained in this section are based on the committee's own study and knowledge of the U.S. forestry research system, on interviews with additional scientists, and on documents the committee received from forestry-associated research organizations. The funding increases recommended in this report reflect the committee's experience in and concern about the current status and future prospects of forestry research in the United States.

Federal support for forestry research has been decreasing over the past decade. The two main sources of support for forestry research are the USDA Forest Service and McIntire-Stennis funds administered through the Cooperative State Research Service. Other federal agencies, such as those within the U.S. Department of the Interior, have minor research commitments. Specifically, the Forest Service budget for research has dropped in buying power by 14 percent over the past 10 years. Competitive grants supported by the Forest Service have been eliminated from their budget. The Cooperative State Research Service will support a modest competitive grants program in forestry ($4 million in 1990), but this remains totally inadequate in both size and scope. Because of inadequate funding, only about 9 percent of the proposals received in 1988 were awarded financial support. McIntire-Stennis funding (formula research funding for state-supported colleges and universities) has been well below the authorization level of 50 percent of the Forest Service research budget. Currently, McIntire-Stennis support is only 13 percent of the Forest Service research budget. Forestry research conducted by industry is also on the decline, with apparently fewer than 12 companies having internal forest biology and management research programs (out of 50 or more companies with substantial forest land holdings). It has been estimated that over the past five years, in-house industry research has decreased by 50 percent. Sources of support for forestry research that seem to be improving include states and private foundations. However, the sum total (including federal, nonfederal, and industry sources of support) of research funding, which is estimated to be about $350 million annually, is very inadequate in view of the societal benefits that can be derived from increased research activity.

In addition to general support for research, the physical facilities and research equipment at most research stations and forestry schools are inadequate. Reports assessing the status of equipment in biology and agriculture research have drawn similar conclusions. These laboratories lack essential resources to carry out state-of-the-art research in forest sciences. Examples of the types of facilities and equipment that need to be available include electron and video-enhanced microscopes, analytical chemistry equipment, biochemistry and molecular biology equipment, computers, geographic information systems, plant growth chambers, and greenhouses. Because funding has not kept pace with changing technology, the technology used in forestry research and teaching is rarely up to date.

Recommendations for increases in funding for forestry research come at a time of overall fiscal constraint for the nation. Government officials must both reduce the national debt and set priorities among competing federal expenditures to enact programs that maintain the welfare, infrastructure, security, and continued economic growth of the United States. As a part of that they must also address public concerns for maintaining global competitiveness and environmental resources. The goal of reducing expenditures while allocating funds for essential programs thus requires fiscal prudence.

The committee recognizes that current federal budgetary constraints make new funds for research support exceedingly difficult to obtain. Meaningful increases in research support for forestry and forestry-related research will likely be realized only as a result of changes in funding priorities within the U.S. Department of Agriculture (USDA) and the U.S. Department of the Interior. As outlined in this report, the need for making these changes in funding priorities is urgent if future forests and related renewable natural resources are to be protected from misuse and environmental degradation and if productivity is to be enhanced.

Conclusions and Recommendations

• Increase competitive grants from the USDA for the five major research areas discussed in this report to $100 million annually. A logical basis for this type of competitive research funding is the research funding initiative suggested by the National Research Council's Board on Agriculture (NRC, 1989c). This proposed increase for forestry research is consistent with the recommendations of the Board on Agriculture, which defines agriculture for the purpose of its $500 million funding initiative to include forestry and related areas. The five major research areas recommended in this report on forestry research (biology of forest organisms; ecosystem function and management; human-forest interactions; wood as a raw material; and international trade, competition, and cooperation) involve areas delineated in five of the six broad program areas in the Board on Agriculture's funding initiative (plant systems; animal systems; natural resources and environment; engineering, products, and processes; and markets, trade, and policy).

• Increase the USDA Forest Service research budget by 10 percent each year for the next five years. These new funds should be allocated to the five research program areas discussed in the report. With these five successive annual increments, the Forest Service research budget will expand from its 1988 level of $135 million to $218 million after five years.

• Increase McIntire-Stennis funds over the next five years to the full authorization level of 50 percent of the Forest Service budget. These new funds should also be allocated to the same five research program areas discussed in the present report. With these five successive annual increments, McIntire-Stennis funding will expand from its 1988 level of $17.5 million to $109 million after five years.

• Conduct a national assessment of the current status of equipment and facilities needed to carry out the research described in this report.

MAXIMIZING THE BENEFITS FROM FORESTRY RESEARCH

Increasing public awareness of the interdependence of our forests and global environmental issues will require greater participation of forest scientists in local, national, and even global policy-making. The demand for expert opinion on environmental issues will increase, as will the demand for expert opinion on human-forest interactions, such as how to manage forests that are increasingly used by people. Therefore, not only is more research necessary, but also more effective ways to transfer this information to policymakers is required.

Extension forestry is an important mechanism for technology transfer, particularly to owners of small parcels of private woodland, natural resource professionals, and the general public. For example, in North Carolina alone, 45 landowner associations have been able to increase timber income by $30 million through effective forestry extension efforts. The present extension forestry effort is inadequate to serve current needs, much less future ones. A strengthened program in forestry research requires a strengthened companion effort in forestry extension.

Conclusions and Recommendations

• Scientists must assume a leadership role in communicating their knowledge to policy-makers.

• Incorporate an outreach component to communicate results of research projects to a broader range of clients whenever possible.

• Establish a professional reward system that acknowledges the validity of efforts of scientists involved in outreach.

• Double the base level of funding and full-time-equivalent staff devoted to forestry extension in cooperation with state and local partners.

• Increase the funding provided by the Renewable Resources Extension Act (RREA) to the appropriation authorization level of $15 million dollars annually.

• Integrate extension specialists with their research counterparts in colleges and universities in those instances where interaction between extension specialists and research scientists is inadequate.

SUMMARY

Forestry research must change radically if it is to help meet national and global needs. It must become broader in its clients, participants, and problems, and at the same time it must both become more rigorous and be carried out in greater depth. The number of scientists and amount of resources devoted to forestry research have declined significantly and are continuing to decline, even as needs increase. To meet the challenge of change, new approaches and new resources of the kind described in this report are required. The educational and fiscal systems that support forestry research must be restructured and revitalized; integrated research facilities must be created where public and private resources can be effectively concentrated on basic questions and extension activities. These changes will be expensive, difficult, and, for many, painful. They will be painful in that research resources will need to be redirected and certain research facilities may have to be closed. The consequences of not making them, however, would be more painful: a national and global society increasingly unable to preserve and manage forest resources for its own benefit and for the benefit of future generations.

We emphasize here that both the wise use and the misuse of forests are consequences of human activity. Without a large additional increment of knowledge derived from increased forestry research to provide policy alternatives, the misuse exemplified by deforestation, destroyed productive potential, and lost biological diversity will prevail. Knowledge gained from an improved system of forestry research will enable society to choose wise use and thus to secure the environmental, economic, and spiritual benefits of forests.

1

Societal Needs and Concerns for the Forest

Concern about the global role and fate of forests has never been greater. Although many opportunities for the use and improvement of forests are apparent, current land-use practices and policies cause unacceptably high damage. Therefore, forestry (see box) and forestry research must become more sensitive to the environmental consequences of land use to provide forest resources for future generations. Some major forest issues that society faces are the following.

• Although the relationship between forests and climate is poorly understood, forests are being rapidly modified or destroyed on a global scale, possibly leading to changed regional or global climates.

• Concern about the loss of biological diversity from forested areas has become global, but management for the maintenance of biological diversity and sustained productivity has been little explored.

• The demand for wood and wood-fiber products to satisfy the basic needs of consumers worldwide has grown by 90 percent in the past three decades; it is expected to grow by another 45 percent by the year 2000 (USDA Forest Service, 1982). Therefore, the United States needs to evaluate its world competitive and cooperative status for supply, demand, and trade of forest products.

• The demands in the United States and elsewhere for increased preservation of "pristine" forested areas and increased recreational use of forests contrasts with the industry view that "the forest products industry faces a timber supply crisis of unprecedented proportions." (American

9

WHAT ARE FORESTS?

WHAT IS FORESTRY?

Forests and related renewable natural resources include the organisms, soil, water, and air associated with timberlands as well as forest-related rangelands, grasslands, brushlands, wetlands and swamps, alpine lands and tundra, deserts, wildlife habitat, and watersheds. These resources include many different categories of land ownership: national forests, parks, and grasslands; federal, state, and private wildlife and wilderness areas; national, state, county, municipal, and community parks and forests; urban and suburban parks and forests; private nonindustrial timber and range lands; and industrial forests and rangelands.

Forestry consists of the principles and practices utilized in the management, use, and enjoyment of forests. Forestry includes a broad range of activities—managing timber, fish, wildlife, range, and watershed; protecting forests and timber products from diseases, insects, and fire; harvesting, transporting, manufacturing, marketing, preserving, and protecting wood and other forest products; maintaining water and air quality; and maintaining society's well-being as it is influenced by forests and other renewable natural resources and their derived products and values.

Forest Council, personal communication, 1989). Even though more timber is currently being grown than is being harvested in the United States, industry faces an acute timber supply problem as a result of new harvesting restrictions.

• Population pressure and economic and social development forecast a growing need to increase forest yields on those lands already dedicated to timber production and to maintain or enhance other outputs (such as recreation, wildlife, and watershed capabilities) of multi-use forest lands. However, our limited understanding of the genetic, physiological, and ecological nature of trees and other forest organisms currently constrains our ability to sustain or increase these yields without risk to forest health.

• The maintenance of forest health in the face of increased regional and global air pollution, altered pest distribution patterns, and human encroachment is a task of high priority if forest cover, watershed values, and wood supply are to be maintained, but the science of integrated forest protection, which would deal with these issues, needs to be developed.

Society's needs and demands from the forest have evolved considerably in the past century. Although wood products still form a necessary component of everyday life, other forest resources, such as wildlife, water, and recreation, are also essential to the quality of life. In addition, increasing concern over environmental degradation (National Research Council, 1986; Wilson, 1988) and global climate change (National Research Council, 1988) is intertwined with these traditional needs and concerns. To be able to meet these needs, new information about forests is required.

Wood accounts for approximately one quarter of the value of major industrial materials. As a result, forests contribute substantially to our nation's economy. Concern about our global competitiveness in the industry is also apparent. In 1987, the United States exported the equivalent of about $8.5 billion in wood products, most in raw wood products such as logs and wood pulp; we imported some $10.9 billion worth of lumber and paper (U.S. Department of Commerce, 1989). Why is the United States exporting timber and importing finished lumber? Questions, such as this, concerning the competitiveness of the forest products industry in the world market have become a critical part of the debate on national competitiveness and productivity.

Forests, however, are more than a place for growing trees for harvest. For example, our urban forests are important to people on a daily basis. Urban forests consist of trees growing in close association with people, such as tree-lined streets and city parks. Urban forests provide valuable shade and protection from wind while enhancing the attractiveness of the places we live and work. Growing trees in an urban setting may offer special advantages in that they not only remove carbon from the atmosphere, they also provide direct shade to humans, reducing ambient temperature and the demand for energy that would otherwise be used for cooling.

Public concern is increasing over the long-term sustainability of all forest resources such as water, fish, wildlife, and recreation. Awareness of the need to preserve biological diversity in forests that serve as reservoirs of plants and animals for ecological restoration is growing. If we are to protect and conserve forest ecosystems, we must embrace a broader concept of the role such systems play in our society and the world and the kinds of education and research programs required within the professional forestry community to provide knowledge for the future. Heightened public awareness about the environment has increased the demand for scientific information to help solve national, regional, and global problems. The public's need for the most up-to-date scientific information for decision-making and public policy has never been greater.

The value and importance of knowledge about forests is increasing, and the kinds of information required differ from those of only a few years ago. Therefore, dramatic changes in forestry research and education are required. Changes and improvements should be made in the basic nature of forestry research, human resources devoted to forestry research, technology and information transfer of the results of forestry research, and the funding of forestry research. In particular, the products of forestry research must be consistent with the land ethic articulated by Leopold (Potter, 1988) as well as in the report *Our Common Future* (World Commission on Environment and Development, 1987). In this regard, forestry research should be a continuing source of information from which improvements in natural resource management can be made more sustainable and environmentally sound.

In this report, we propose a research agenda for forestry that will result in new and innovative approaches to arrest forest ecosystem degradation and provide the wide array of needed forest products. By gaining an understanding of the fundamental structures, processes, and relationships within forests, we will be better able to anticipate responses, management needs, and capabilities of forest ecosystems to buffer changes in the environment. By understanding the effects of humans on the forest, we will be better able to meet our needs and promote our long-term well-being. We will use this scientific information to design strategies for sustaining forest ecosystems and their products and to make public policy decisions to meet these strategies.

2

The Gap Between Society's Needs and the Status of Forestry Research

Research on forests and their components and their interactions with people currently spreads far outside forestry institutions and companies. No scholarly effort to fully inventory research on forests has been made in the United States, nor is there any widely recognized forum that facilitates communication among all the parts of forestry and forest research. This lack of leadership contributes to an overall fragmentation of effort and absence of clear definition of what constitutes forestry and forestry research. This fragmentation has its ultimate roots in the various concepts of forest management as they developed in this country in the late nineteenth and early twentieth centuries.

During the past two decades, natural resource management in general and forest management in particular have been in a state of considerable turmoil. While resource managers have been struggling with new views and values, forestry research has concentrated primarily on technical forestry or production-based forestry. This tendency, coupled with the public's increasing interest in environmental issues, has led to a widening gap between the status of forestry research and the perceived need for this research.

To form a vision of future forestry research needs, we must understand our current situation. Throughout most of the nineteenth century, natural resource use in North America was based on a utilitarian paradigm, which held that natural resources were inexhaustible and that they should be exploited to raise individual and collective standards of living.

New views on how natural resources should be used began to develop in the latter half of the nineteenth century. George Perkins Marsh (1864) was an early American proponent of the conservationist paradigm. Marsh urged that humanity address nature with a sense of stewardship. It was logical that the progressive administration of Theodore Roosevelt should adopt this philosophy for natural resources, and Gifford Pinchot, as a member of this administration, became not only the father of American forestry, but also the acknowledged father of forestry's conservation paradigm (Culhane, 1981).

Pinchot transformed the philosophy into one of wise use; he championed scientific forestry and rational planning as ways of using forests to raise living standards without destroying the land's ability to be used. Pinchot's vision was, however, still biased toward commodity production from the forest, which he referred to as a "wood factory" and a "tree farm."

This conservation paradigm had insufficient public support to enable the Roosevelt administration to persuade Congress to enact conservation legislation. Not until Pinchot's ideas were blended with the more romantic views of others such as John Muir and Frederick Law Olmstead was public sentiment sufficient to force Congress to act. Consequently, public support for "conservation" has always come partly from those dedicated to conservationism and partly from those who believe deeply in more romantic views of nature (preservationists) (Hays, 1959).

This dichotomy of attitudes toward public land management among the citizenry has been a major factor in the turmoil over forest plans, oil and gas leasing, grazing fees, and so forth. With a few notable exceptions, such as the spotted owl, research directed toward the major concerns of the preservationists had been a low priority. Foresters inherited Pinchot's "tree-farm" view of the forest and his belief in scientific forestry as the road to wise use of forestry resources. Research priorities have long been dominated by commodity production goals.

Aldo Leopold expressed a divergent view of natural resources management in "The Land Ethic" (Leopold, 1949). In addition to urging "an intense consciousness of land," he provides a dramatic test for land-management activities: "A thing is right when it tends to preserve the integrity, stability, and beauty of the biotic community. It is wrong when it tends otherwise."

In the last decade of this century, we see a renewed emphasis on a land ethic, but this time with a global perspective. This has resulted in renewed support for an environmentalist paradigm. Such a model of resource management is perhaps best described in *Global Bioethics* (Potter, 1988), *Our Common Future* (The World Commission on Environment and Development, 1987), and *The Closing Circle* (Commoner, 1971). It holds that human beings and nature are interrelated, that humans are not superior to the natural world, but depend on the biosphere for their existence. The biosphere's resources are finite, and human activities must not destroy the biosphere's intricate workings.

Human intervention into the global flows of energy and matter now equal or exceed the magnitude of natural processes (NRC, 1988). Soil acidification, erosion, and declining production yields in several regions call to question the foresters' view of forests as renewable plantings. Simultaneously, the continued decrease in biological diversity in our parks and preserves challenges the effectiveness of the set-aside approach to biological diversity conservation (National Parks and Conservation Association, 1989). Simply setting areas of land aside for conservation and preservation will be inadequate to ensure the survival of certain species. Sustainability of forest production systems is of utmost importance, as is the maintenance of our biological diversity, environment, and aesthetic resources. Thus, the traditional models for natural resource management are inadequate.

The new environmental paradigm will demand more of science than did its predecessors. The kind of forestry research we have been conducting will need to continue, but research priorities must be much broader. The breadth of forestry and the study of forest resources requires information and expertise that must include principles of basic biology, ecology, agriculture, forest management, engineering, sociology, and economics. As an emerging discipline, conservation biology testifies to the need for traditional institutions to respond more rapidly and effectively to the needs and challenges posed by those trying to bridge the gap between basic biology and applied research.

An examination of the relationship between forestry and traditional agriculture is also informative. The most important similarities between forestry and agriculture are in our shared scientific roots—botany, zoology, soil science, genetics, mycology, plant physiology and pathology, entomology, ecology, microbiology, and statistics. Silviculture is the scientific counterpart of horticulture or agronomy. Certain specialized fields of forestry, such as wildlife and fisheries biology, forest engineering, and forest economics, have profited from related scientific endeavors in animal husbandry, agricultural engineering, and agricultural economics. The flow of ideas between agriculture and forestry has been generally from agriculture to forestry, largely because of the much larger investments in agricultural research.

But critical differences exist between forestry and agriculture. Forests must be managed over very long periods of time (long rotation), whereas many agricultural products are harvested annually (short rotation). Because of the vast size of forests and their natural state, management necessarily tends to be extensive rather than highly intensive, as agricultural production is. Forestry generally manages for an array of different products and values simultaneously; in agriculture, relatively few products are produced in a given area. Whereas agricultural fields are of human creation, most forests are essentially developed by natural systems. These natural processes are highly complex and can be very different from one forest area to the next.

Over the past decade, both agriculture and forestry have been stressing the need to move to more environmentally sound (and therefore more sustainable) production practices for agriculture (NRC, 1989a) and management practices for forestry. With this common goal of sustainability, scientists doing research in agriculture and forestry must interact more than in the past.

FUNDING FOR FORESTRY RESEARCH

In sharp contrast to the growing public perception of forests' importance and real societal needs, forestry research is meagre (Mergen *et al.*, 1988) and declining in most sectors (Giese, 1988). Funding is one measure of research vitality. To what extent is the forestry research enterprise being supported?

Federal Sources of Support

Federal research support comes primarily from the activities of the U.S. Department of Agriculture's (USDA's) Forest Service (Chapman and Milliken, 1988) and secondarily from the McIntire-Stennis Act administered through the USDA's Cooperative State Research Service (CSRS). These two sources usually account for more than 80 percent of the total federal expenditure specifically designated for forestry` research. Other federal sources providing support for forestry research include the National Science Foundation (NSF), Hatch funds (which provide support to state agricultural experiment stations administered through CSRS), the U.S. Department of Energy, the U.S. Department of the Interior, and the U.S. Environmental Protection Agency.

USDA Forest Service. The Forest Service research budget has been steadily declining in purchasing power (Table 2-1). Concomitantly, the amount of Forest Service support for extramural programs (money provided by the Forest Service to non–Forest Service research) has also decreased. As a direct result of these budget restrictions, the Forest Service has been forced not only to reduce the size of its scientific staff by 25 percent, but also to reduce the number of research facilities by 12 percent and research work units by 23 percent since 1978 (Table 2-1).

Not all areas in the Forest Service research budget are decreasing, however. The five main Forest Service forest research budget items are: forest protection, resource analysis, timber management, forest environment research, and forest products and harvesting. Within these budget

TABLE 2-1. Forestry Research Funding Statistics for the USDA Forest Service, 1977-88.

Year	Appropriations (Million $)		Extramural funding (Million $)		Scientist-years	Research locations	Research work units
	Actual	1982	Actual	1982			
1978	88.4	122.1	11.0	15.2	962	86	247
1979	93.9	120.8	10.0	12.9	942	86	245
1980	95.9	111.3	10.6	12.3	964	86	248
1981	108.5	116.2	14.2	15.2	958	85	242
1982	112.1	112.1	10.8	10.8	908	83	235
1983	107.7	102.9	9.3	8.9	838	80	219
1984	109.4	99.5	7.7	7.0	813	76	207
1985	121.7	105.9	7.5	6.5	799	76	200
1986	120.1	101.5	10.4	8.8	734	76	199
1987	132.7	107.7	14.6	11.9	713	76	200
1988	135.5	105.2	18.3	14.2	724	76	190

SOURCE: B. Weber, USDA Forest Service, Washington, D.C., personal communication, 1989.

items are six priority areas: global climate change, catastrophic forest fires, water quality, expanding economic opportunity through new wood products, southern forest productivity, and threatened and endangered species. Each of these priority areas is expected to receive an increase in fiscal year (FY) 1990 over FY-1989 levels ranging from 35 percent for water quality research to 3 percent for research on catastrophic forest fires (USDA Forest Service, 1989).

Compared with other federally funded research programs, the Forest Service research budget is very small. For example, in 1988 the USDA's Agricultural Research Service (an agency devoted to agricultural research) was funded at about $541 million, while Forest Service research received only about $135 million.

Competitive Research Grants. One important source of funds that has come from the Forest Service budget is the competitive grants program administered by the Competitive Research Grants Office (CRGO) of CSRS. From 1985 until 1988, the Forest Service provided funds to CRGO to support grants in forest biology and wood products research on a competitive basis. Funding decreased from $7.5 million in 1985 to $3 million in 1988. In 1989, no funds were provided for this program. Through the Agriculture Appropriations Subcommittees, which act on the CSRS budget, this program will receive renewed funding at a level of approximately $4 million in FY-1990. In 1988, a consensus statement was developed that recommended the support of competitive grants in forestry research (published jointly by the Forest Service, the American Forest Council, and the National Association of Professional Forestry Schools and Colleges). Groups involved in this consensus statement included trade associations, professional societies, and government agencies.

In FY-1988, 235 proposals were received by the CRGO forestry program: 120 in forest biology and 115 in wood utilization. Although about half of all research proposals received were judged to have some merit by peer review, only 13 were funded in wood utilization and 9 in forest biology (only 20 percent of all meritorious proposals were funded) (CSRS, 1989).

In addition to the CRGO forestry program, other programs within CRGO and NSF support some forestry-related research on a competitive basis. The total amount of competitive support for forestry research is less than $5 million annually. When this amount of competitive support is compared with the research needs described in Chapter 3, it is apparent that increases in research support are imperative. A substantial source of forestry-related research support comes from the NSF's Division of Biotic Systems and Resources (total budget of about $60 million in 1988). Much of the research supported by this division could be applied to forestry research if scientists supported by NSF and scientists in traditional forestry disciplines interacted more.

McIntire-Stennis Funds. The McIntire-Stennis Act of 1962 provides financial support (based in part on matching funds) to public colleges and universities with forestry research and graduate programs for the long-term studies essential to advances in forest productivity. In addition to supporting scientists in forestry schools and colleges, McIntire-Stennis funds should be used to encourage other scientists to initiate forestry research.

Currently, three major factors determine the proportion of funds a state will receive from this program. The first factor, weighted at 40 percent, is the proportion of acreage in commercial forest land; the second, also weighted at 40 percent, is the volume of roundwood produced; and the third, weighted at 20 percent, is the amount of nonfederal money spent on forest research. A certain amount of flexibility exists in this formula; for example, the weight of the factors is not mandated by law, but is set by the Secretary of Agriculture. A reevaluation of this formula based on the current status of forestry research would be appropriate.

The McIntire-Stennis funding program is an effective leveraging scheme by which, for each federal dollar provided, five to six nonfederal dollars are spent (B. Post, CSRS, personal communication, 1989). Except as indexed by matching dollars in forestry research, this mechanism was not designed to explicitly consider the research capability or quality of research programs at the participating institutions.

McIntire-Stennis funds are not allowed to exceed 50 percent of the funds appropriated for forestry research for the Forest Service. And yet the appropriation is only 13 percent of the support provided to the Forest Service. Table 2-2 depicts the steady erosion of the buying power of McIntire-Stennis funds due to inflation. As seen in Table 2-3, the support provided by McIntire-Stennis for university research has been decreasing in purchasing power, from $13 million in annual expenditures in 1978 to only $11.9 million in 1988 (both in 1982 dollars). This situation, however, appears to be improving, with $17.5 million appropriated in 1988 (Table 2-2). (It is important to realize that budget numbers given in Table 2-2 are actual fiscal-year appropriations passed into law by Congress. Table 2-3, however, contains budget information on university expenditures that may cover money appropriated over several fiscal years.) In 1987, the National Association of Professional Forestry Schools and Colleges recommended that McIntire-Stennis funding be increased to $25 million. The forest products industry through the American Forest Council (AFC) also endorsed this recommendation (USDA, 1987). Approximately 40 percent of the forestry research expenditures reported by CSRS goes to forest biology research compared with about 15 to 20 percent for forest products research (B. Post, CSRS, personal communication, 1989).

Another source of formula funds is the Hatch Act, which is used to support research to promote sound and prosperous agriculture and rural life. In FY-88, approximately $147 million of Hatch Act funds were used to support research [as reported to the Current Research Information System

TABLE 2-2. McIntire-Stennis Fund Appropriations, 1978-1988.

Year	Appropriations (Million $)	
	Actual	1982
1978	9.5	13.1
1979	9.5	12.2
1980	10.0	11.6
1981	10.8	11.5
1982	12.0	12.0
1983	12.5	11.9
1984	12.7	11.6
1985	13.1	11.4
1986	12.4	10.5
1987	12.4	10.1
1988	17.5	13.6

SOURCE: W. Murphey, USDA Cooperative State Research Service, Washington, D.C., personal communication, 1989.

TABLE 2-3. Source of Annual Expenditures on Forestry Research at Universities, 1977-88 (in millions of dollars).

Year	McIntire-Stennis		Total federal		Nonfederal		Federal percent of total (1982)
	Actual	1982	Actual	1982	Actual	1982	
1978	9.4	13.0	21.4	30.0	30.1	41.6	42
1979	9.3	12.0	22.4	28.8	34.2	44.0	40
1980	9.5	11.0	25.4	29.5	41.9	48.6	38
1981	10.1	10.8	27.9	29.9	45.5	48.7	38
1982	11.0	11.0	28.4	28.4	48.4	48.4	37
1983	11.4	10.9	26.7	25.5	46.7	44.6	37
1984	11.8	10.7	25.3	23.0	49.5	45.0	34
1985	11.8	10.3	25.9	22.5	55.4	48.2	32
1986	11.4	9.6	30.7	26.0	65.0	54.9	32
1987	11.4	9.3	28.5	23.1	65.9	53.5	30
1988	15.3	11.9	36.2	28.1	70.3	54.6	34

SOURCE: W. Murphey, USDA Cooperative State Research Service, Washington, D.C., personal communication, 1989.

(CRIS)], of which only $3.2 million went to support forestry research (W. Murphey, CSRS, personal communication, 1990).

State Sources of Support

Individual states are the major supporters of forestry research at the nonfederal forestry research institutions, which are predominantly members of the National Association of Professional Forestry Schools and Colleges. State funding currently constitutes about two thirds of total university support for forestry research and has increased slightly in response to decreased federal support. State funding accounted for about 12 percent of all forestry research support in 1977 and increased to 20 percent in 1986 (Giese, 1988). A summary of support for forestry research at universities expressed as annual expenditures in various categories reported through CRIS can be seen in Figure 2-1.

Industrial Sources of Support

The extent of industrial forestry research is difficult to estimate accurately. Estimates generally range between $50 million and $100 million per year, but these do not include funds for products research (American Forest Council Research Committee, personal communication, 1989). It is clear, however, that few companies support internal research programs, especially forest biology research. Fewer than 12 companies have internal

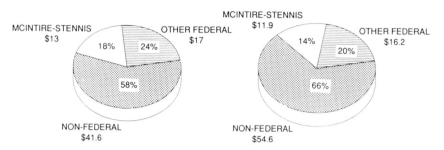

FIGURE 2-1 Annual expenditures on university forestry programs in millions of constant 1982 dollars as reported to the Current Research Information System (CRIS). (Constructed from data in Table 2-3.)

biology research programs (R. Slocum, North Carolina Forestry Association, personal communication, 1987). It has been estimated that in-house industry research programs have decreased by 50 percent over the past five years (Giese, 1988). However, industry has been active in forming university-industry research cooperatives (AFC, 1987). Survey results published in 1987 indicate that of the 51 responding schools, 19 had research cooperatives with industry. These cooperatives have $5.2 million in funding, of which industry provides $2.5 million (AFC, 1987).

Private Universities and Foundations as Funding Sources

Other sources of research support are private universities and foundations. In 1987 and early 1988, approximately $134 million was spent on environmental law, protection, and education (Foundation Center, 1989). Of this $134 million, approximately $27 million is characterized by the Foundation Center as being spent on research. Much of this money, however, is not spent directly on research that can be classified as academic research— that is, research that might be associated with a university. Approximately 80 grants from private foundations listed forestry as a specific subject area, and more than 200 grants listed natural resources conservation. Examples of major supporters of this research include the Ford Foundation, the William and Flora Hewlett Foundation, the W. Alton Jones Foundation, the MacArthur Foundation, the Andrew Mellon Foundation, and the J. N. Pew, Jr., Charitable Trust.

Support for forestry programs at private universities, such as those at Duke, Harvard, and Yale, comes predominantly from nonfederal (such as private foundations) and competitive sources and from endowment funds.

Organization of Research Funding

Major changes are needed in the organization of research funding to facilitate creative research. At present, formula funding through state or federal sources or through the Forest Service provides most of the funds for public research. Although much of this support is needed for long-term programs, significant increases are needed for new programs that are investigator-initiated, such as peer-reviewed competitive grants. Peer review provides a necessary incentive for research that is novel, competent, and creative.

Currently barriers stand before more effective research. Examples of these barriers include the following.

• Forest scientists working for the federal government are not allowed to apply for competitive grants from certain other agencies, such as the National Science Foundation; scientists working for any agency should be allowed to compete on merit for competitive funds from any agency.

• Forestry research is sometimes not included in other basic science funding programs; where the research areas are appropriate, forestry research should not be excluded.

• Nonforestry scientists have little opportunity to compete for funds that are currently awarded on a noncompetitive basis to forestry scientists; what is needed is a more open and competitive approach that would provide funds previously available to traditional forest scientists as well as provide more access to nonforestry scientists to work on forestry-related problems.

Extension Forestry

Forestry extension is the primary mechanism in the United States for technology transfer from research programs to users—traditionally private forestland owners. As with other extension programs, forestry extension is a cooperative activity among the federal government, states, and counties. The relationships among the various cooperators in forestry research and extension have recently been examined (Rogers *et al.*, 1988). This report recommends major changes in the way universities, government agencies, and other organizations interact to improve technology transfer in renewable resources. Forestry extension is supported, in part, by the federal government through the activities of the Smith-Lever Act, which provides formula funding to states for all extension activities, and through the Renewable Resources Extension Act (RREA), which targets money specifically for extension efforts on renewable resources including forestry. In 1988, the total amount spent by the federal government on extension in natural resources programs was only $9.7 million, compared with $345 million in the total extension budget (D. E. Nelson, Extension Service, personal communication, 1989). In 1988, RREA received only $2.8 million of the $15 million authorized under the RREA legislation (D. E. Nelson, Extension Service, personal communication, 1989). In addition to direct support, the funds provided through RREA are effective at leveraging about $2 dollars from nonfederal sources for every dollar spent by the federal government (USDA Extension Service, 1986).

In 1985, 573 full-time equivalents (FTEs) were in the cooperative extension system's natural resources programs (USDA Extension Service, 1986). Of these, 258 were in forestland management, 79 in rangeland management, 131 in wildlife and fisheries management, 39 in outdoor recreation, and 66 in environmental quality and public policy. In just the forestland management category, an estimated 200 FTEs need to be added to address future extension needs adequately (Backiel, 1986). An extension area that requires special attention is urban forestry. On the basis of the latest data available (1979), only 24 states had extension programs in urban forestry, and only about 19 FTEs serviced those programs (Science and Education Administration, 1980). In addition, new challenges posed by potential environmental changes present forestry extension with a wide array of new areas that need to be addressed, such as forest management under potentially changing climatic conditions and the preservation of biological diversity. Incorporating advances made through biotechnology will also require new approaches and special attention—for example, public education on the potential benefits and risks posed by introducing new genetically modified organisms into forest ecosystems. The future of forestry extension is clearly broader and more complex than its past. Many of these activities will require a host of new experts and extension specialists as well as creative new methods of technology transfer and public education.

THE STATUS OF FORESTRY EDUCATION AND INTELLECTUAL LEADERSHIP

The Production of Ph.D.s

Since the late 1970s, the number of students earning Ph.D.s in forestry has not increased significantly. From 1984 to 1988, an average of 98 Ph.D.s were granted each year, up slightly from an average of 86 per year between 1979 and 1982 (NRC, 1989d). Of the 106 Ph.D.s awarded in 1988, 13 were awarded to women and 24 were awarded to non–U.S. citizens with temporary residency status. Of the 82 Ph.D.s awarded to U.S. citizens, only 5 were awarded to members of minority groups. Areas allied to forestry include fisheries science, wildlife management, and ecology, where 42, 39, and 155 Ph.D.s were awarded in 1988 (NRC, 1989d). Undergraduate degrees in forestry have decreased to approximately 50 percent of their number in the late 1970s (Society of American Foresters, 1988).

When academic and nonacademic forestry groups were surveyed to determine the major issues in forestry education, the following were regarded as most important (Forestry Education Problem-Assessment Steering Committee, 1987):

- the need to attract high-quality students,
- the need to maintain and improve teaching quality,
- the need to move away from technician-type courses to more analytical, decision-making, conflict-resolution types of courses,
- reduced employment opportunities,
- the lack of a clear definition of the forester's role.

These issues are clearly interrelated. For example, to attract more students of high quality, attention must be paid to teaching quality, curriculum issues, and future employment. Perhaps the most important issue is the need for a clearer definition of the mission of colleges and departments of forestry. An important objective for those concerned with forest science should be the establishment of an intellectual environment that fosters excellence, provides rigorous education, and attracts specialists from nonforestry backgrounds. With a clearer definition, the rest of the issues relating to forestry education can be adequately addressed.

To acquire the necessary information base through research will require an increase in intellectual resources devoted to forestry research. This increase should occur within the next five years if research needs are to be adequately met. Because of funding constraints, the forest science community does not now have the human resources to do the research our nation requires. Thus, a great need exists for increased numbers of forest scientists as well as for a change in their education. This increase could be accomplished, in part, by recruiting and educating scientists from groups that are traditionally underrepresented in forestry research, such as women and members of minority groups.

One approach to rapidly increasing intellectual resources is to promote interdisciplinary research. Research programs that recruit scientists from related disciplines provide new technology and different research approaches. One such opportunity lies in increasing the interaction of traditional forest biology with molecular biology. Traditional forest biology has a large number of well-defined problems, such as understanding and manipulating disease resistance genes, that would be advanced significantly if it could integrate the technology of plant molecular biology with the forest sciences. Other examples of needed research integration are in the areas of atmosphere-biosphere interactions, mathematical modeling, and ecosystem science.

Intellectual Leadership in Conservation

Over the past several decades, the forestry profession has lost its leadership role in the conservation movement in this country. In the first half of this century, many of the leaders in the conservation movement were professional foresters and many were researchers. They not only created a vision based on "the greatest good for the greatest number in the long run," they proposed that the vision be met through "scientific forestry." They persuaded the public of the validity of their approach.

Today few public opinion leaders who help shape policy on natural resources are professional foresters, and even fewer are researchers. While foresters are accused of having sold out to commercial interests, others (often from narrow special-interest groups) are leading in the reshaping of the conservation movement and of forest-related policy.

If forestry research is to remain relevant to the conservation and management of natural resources, forestry researchers must reassume some of the leadership of the conservation movement. They must regain public trust and strive to educate the public so that it may wisely influence the formation of forestry and natural resources policy. The power of modern science and technology can be brought to bear on natural resource issues only if policy allows it. This is unlikely to occur unless the forestry profession exerts leadership.

Employment

Approximately 50 percent of new Ph.D.s in the forest sciences are employed by universities, while 6 and 7 percent are employed by the Forest Service and industry, respectively (Forestry Education Problem-Assessment Steering Committee, 1987). Between 1978 and 1988, the number of scientist-years in the Forest Service has been reduced by 25 percent—from 962 to 724 scientist-years (Table 2-1). During that time, university scientist-years have decreased by roughly 8 percent (Giese, 1988). In general, the demand for forestry Ph.D.s in government, industry, and universities is slight. In addition to forestry Ph.D.s, scientists without traditional forestry backgrounds are being hired to fill research positions. However, the demand for these scientists is also low in the forestry research community.

This low level of employment demand does not reflect the need for forestry research. The committee believes that the decrease in demand for forestry Ph.D.s is a symptom of short-sighted funding perspectives that fail to address many of society's needs adequately. In fact, even if the recommendations of this report are implemented, the current production of forestry research scientists will remain inadequate. Furthermore, the need will increase for Ph.D.s with expertise in areas such as geographic information systems (GIS), molecular biology, forest policy, ecology, and landscape management, to name a few. In addition, scientists with an international perspective—especially on tropical regions—will be in demand. It is important to keep in mind that the quality of these new scientists will need to be high to meet the challenges that forestry research is facing now and will face in the future.

3

Areas of Research

Despite the excellence of some individual researchers and centers of forestry research, the quantity and coherence of research do not match even current needs. Changes in public perceptions and uses of forests call for new information not now provided in sufficient depth. Not only is an increase in depth and quantity needed, but the quality of this research must also be high. Unprecedented opportunities to understand forest structure and function presented by ecosystem ecology and molecular biology are not being pursued with enough talent and vigor. Opportunities also exist to better understand human-forest interactions, economics and forest policy, and wood as a raw material. Increased funding is recommended in the following five major areas: (1) biology of forest organisms, (2) ecosystem function and management, (3) human-forest interactions, (4) wood as a raw material, and (5) international trade, competition, and cooperation. These five areas represent the broad cross-section of forestry research seen by the committee as central to addressing present and future societal concerns pertaining to forests and many global environmental issues. Most of the research needs described in this chapter have importance for tropical as well as temperate-zone forestry. Rather than divide research needs into "tropical" and "temperate" categories, we intend to highlight fundamental research areas that need to be strengthened. Research needs specific to each section discussed in this chapter are listed in Appendix B.

BIOLOGY OF FOREST ORGANISMS

Forests are inhabited by thousands of noncommercial animal, plant, and microbe species that coexist with commercial tree species. Many of these organisms perform for the commercially important species critical ecological functions such as nitrogen fixation, pollination, propagule dispersal, mitigation of forest pests, and enhancement of nutrient uptake. Forest management activities affect the habitats of all these organisms. Our understanding of ecosystem function and our ability to improve forest productivity, to ameliorate the effects of environmental disturbances, and to restore ecosystems are hindered by our limited knowledge of the basic biology of forest organisms. Thus, research not only on trees, but also on other plants, animals, and microbes presents opportunities to advance science and better the human condition.

Physiological and Genetic Bases of the Mechanisms Underlying Forest Health, Productivity, and Adaptability

Forest environments are changing at an accelerating rate; yet, for the most part, we do not know how forest organisms will adapt and respond to these changes. Societal demands on forests also continue to increase, and we must gain fundamental knowledge to allow development of cultural practices and tree varieties or genotypes that sustain and enhance productivity. These issues require fundamental understanding of the physiological and genetic mechanisms controlling the growth and development of key forest organisms and the interactions occurring among these organisms. This information is needed, for example, to understand and predict how tree genotypes will respond to environmental stresses (such as air pollution and a warming climate) and to identify traits that are particularly beneficial or useful in specific situations.

Selected microbes, insects, animals, and forest plants other than trees are important for forest sustainability or the maintenance of biological diversity. Basic physiological and genetic information about them is crucial. By providing a physiological and genetic understanding of the mechanisms governing adaptive responses of these organisms, we will have a solid basis for designing cultural practices that will maximize sustainability, productivity, and survival.

The great diversity of forest organisms and the variety of their interactions makes it difficult to narrow the selection of specific organisms or processes that merit in-depth genetic and physiological analyses. However, concentrated analyses of a few model systems have been essential for recent progress in biology, and the selection of a few key forest organisms could enhance forest research. Already established and easily manipulable biological model systems such as *Drosophila* or *Arabidopsis* can be used by forest researchers to identify genes or gene functions. Information obtained from these established systems can be directly or indirectly applied to numerous forest organisms. Ultimately, however, the mechanisms or processes must be verified and characterized in the forest organisms of interest themselves.

Molecular Genetics of Forest Organisms. Forest trees have been difficult objects for genetic studies because of their large size and long generation times. The new technology of recombinant DNA makes it possible to study the genetic structure of organisms and populations in detail. These techniques provide new opportunities to study the molecular bases of a wide variety of genetically regulated processes in growth, metabolism, development, response to environmental stimuli, and evolution. These methods can contribute to the understanding of the mechanisms that underlie host-pathogen interactions, the responses of forest trees to pests and abiotic stress, and interactions with other forest organisms.

Studies of the molecular interactions of hosts and pathogens can help to identify specific genes involved in the resistance to disease or in other aspects of defense response. Breeding and selection of certain forest tree species have produced lines that are resistant to major diseases such as fusiform rust and white pine blister rust. Although progress has been slow, good material is available and the promise for the future is bright if the techniques of molecular biology can be applied. Similarly, the technology can be applied to the study of the genetic properties of pathogens and pests themselves to provide information that could lead to the development of better biological control agents (see Pest Management).

Molecular Markers for Stress Responses in Forests. Much concern has been expressed about the health of our forests in the face of environmental pollution and the prospects of global change, including changing climate. Little is known about the molecular physiology of stress responses in forest trees and other organisms. If the molecular events in stress responses become better understood, it will be possible to identify the key factors in the responses of forests to environmental stress. In this way, it will become possible to monitor the health of forest trees, diagnose problems, determine the biological bases for declines in specific forests, and take remedial action to improve forest health.

Genetic Markers. In agronomic crop plants it is generally accepted that genetic markers will be a significant aid to improved breeding by early selection of desirable traits. The useful genetic markers called restriction fragment length polymorphisms (RFLPs) are based on the establishment of a specific set of DNA probes obtained by recombinant DNA techniques.

These markers are used to establish detailed genetic maps of important species that permit accelerated breeding strategies. Breeding can be accelerated when linkage between genetic markers and important traits can be determined.

Once detailed maps are established, genes that confer useful properties —such as those for improved wood properties, disease resistance, or stress tolerance—can be located and identified. More importantly, the technology provides the opportunity to begin to dissect quantitative traits into genetic factors that could have major effects on specific phenotypes, but that have not been analyzed because appropriate phenotypic markers have been absent. Detailed genetic markers will also be of major value in the identification of relationships of diverse genetic material, whether in testing of seed sources or in establishing the interrelationships of organisms at the population level.

Genetic markers and genetic maps will also be useful in studying population genetics and evolution of forest organisms. For example, monitoring gene frequencies in a given population can provide valuable information on the effects of a shrinking population as a result of reduced habitat. In addition, genetic markers can be used to shed light on the behavioral ecology of forest organisms. In particular, an organism's behavior can now be accurately linked to its reproductive success through genetic profiling or fingerprinting.

Genetic Engineering. Genetic engineering of forest organisms will make possible the use of genetic material that is not currently in the natural breeding population of a particular forest species. For trees, it makes possible the use of genetic material from other plants, microbes, and animals that could confer resistance to diseases, pests, and toxic chemicals, as well as the ability to detoxify pollutants, respond to environmental stresses, and a wide variety of other characteristics that might not exist in the native species. To make genetic engineering of trees possible, new technology must be developed and much basic information must be obtained about the molecular processes that may need to be modified. Other forest organisms might be engineered to improve biocontrol agents, nutrient uptake, and pollution control. We must also continuously evaluate potential hazards that might arise from the large-scale use of genetically engineered organisms.

To maximize the usefulness of genetic engineering, genes must be preserved and studied. The best way to do this is through the preservation of biological diversity. The loss of biological diversity through habitat destruction, together with its other undesirable consequences, may cause existing gene pools to be inadequate for species improvement by genetic engineering.

Cell and Tissue Culture of Forest Trees. Cell and tissue culture continue to play important roles in the cellular biology of forest trees and in micropropagation for commercial forestry. Limitations of the methodology of tissue culture constitute the major research barrier to the development of DNA transfer methods that are needed for genetic engineering. Greater effort is needed to overcome the problems of tissue-culturing many forest tree species. In addition to its usefulness in genetic engineering, tissue culture can provide methods for investigating biochemical, physiological, and developmental characteristics of plant cells that could not be studied in the intact plant.

Evaluating the Potential Risks of Genetically Modified Organisms in Forests. Applications of biotechnology in forestry require insights into natural ecosystems (NRC, 1989b). Organisms that are genetically modified need to be evaluated to ensure that they pose no unacceptable risk to the environment or to human health. Emphasis should be given to evaluating the impacts of the unique properties of modified versus unmodified organisms. Considerable research will be required if we are to learn about the biological processes and interrelationships within forests so that the safety of modified organisms can be evaluated responsibly. A great deal of work is and will continue to be needed to educate the public on both the value and the safety of the new technologies. In addition, forest scientists will be needed to assist regulatory agencies in developing and evaluating scientifically sound guidelines for monitoring the release of genetically engineered organisms in agriculture and forestry so that our forests will be adequately protected.

Biotechnology and the Forest Environment. Biotechnology has the potential to help solve many problems of forest environments through treatment of waste products of the forest products industry, improved efficiency of growth and processing of trees, and the use of molecular technology in monitoring the responses of forests to environmental stress. For example, lignin-degrading enzymes from fungi or isolated genes and their products might be used to degrade lignin wastes and reduce the impact of pulp mills on local environments (Tien and Kirk, 1983). Similarly, biotechnology may improve the efficiency of tree growth by modifying the trees themselves or their symbiotic microbes. In addition, the impact on the environment of growing trees for wood in intensive wood-production systems could be reduced if the properties of wood could be modified to increase yields through more efficient processing. In this way, less land could produce the same amount of wood so that the impact on local or regional environments would diminish (see Wood as a Raw Material).

Long-Term Site Productivity

Sustainable productivity is an emerging scientific topic in the fields of natural resources and agriculture. Soil, pests, and environmental change and degradation have escalated concerns about our ability to maintain current yields of food and fiber (let alone to increase yield). These concerns have clarified our inadequate knowledge of the basis of productivity. National concern exists about the relationship between forest management activities and timber production, especially harvesting and site preparation, and about the capacity of the land to generate other benefits on a sustained basis. Our lack of understanding of the biology, culture, and protection of urban forests is also severely limiting management of those forests and policy-making for our future use and enjoyment of them.

Pest Management

Agricultural scientists have shown that one of the most efficient, long-lasting solutions to disease and insect problems is the use of resistant varieties. Disease- and pest-resistant trees have been developed through sexual hybridization and selection, but this is a slow and inefficient process for tree improvement. Forest scientists need to increase the pace of their search for resistance to major tree diseases and insect pests. New biotechnology methods need to be developed as tools for identifying and utilizing new sources of resistance.

Integrated pest management is an ecological approach that utilizes biologically effective, environmentally safe, and economically sound methods of managing pests. Although it is not a "new" scientific issue, it is receiving increasing attention as practical problems and social concerns constrain traditional chemical approaches to pest control. The use of insecticides, fungicides, herbicides, and biological control are all involved in the development of a comprehensive strategy for minimizing pest problems. The desire for environmentally sound methods of pest management is intensifying, and the philosophy underlying the chemical targeting of organisms is increasingly questioned. The development of integrated pest management methods for forest resources is a major scientific challenge.

Biological Pest Control. Agriculture has experienced problems over the past years in the utilization of broad-range chemicals for the control of various plant pests. Foresters would thus be wise to anticipate the many ecological and practical problems that stem from the use of fungicides, herbicides, and insecticides for the control of forest pests and to minimize the use of such chemicals in forest management. Biotechnology will provide opportunities for alternative approaches to the control of pest populations that are generally more specific and therefore less environmentally disruptive than broad-range chemical approaches. For example, some viruses reduce gypsy moth populations specifically while being safe for other insects and other forest inhabitants (including humans). Such viruses are being developed for the control of several key forest insect pests, but much more information is needed on these and other microorganisms (bacteria, fungi, protozoa, viruses, and so forth) affecting other forest pests. The effective utilization of such microbial pest control methods requires an intimate understanding of the nature of these microorganisms and their interactions with target and nontarget pests.

In considering biotechnological solutions to pest control, biological control solutions other than the introduction of pathogens, parasites, and predators should also be pursued. Many of these additional biological control solutions are based on such biological strategies as biochemical signals utilized by the pests (pheromones, allemones, and others) and biochemical mechanisms of natural plant resistance to pests. Basic research will be needed to identify such signals and determine how they can be incorporated into pest control strategies.

ECOSYSTEM FUNCTION AND MANAGEMENT

Many of the emerging issues in science have a great impact on societal problems, especially problems in the area of natural resources and environment. In effect, the old distinction between basic and applied sciences is being dissolved, and scientific agendas are evolving around critical environmental issues such as the effects of global warming and of loss of the ozone layer and biological diversity. Hence, the scientific issues identified in this section are strongly linked to issues of resource management. Much scientific effort, however, needs to be devoted to developing basic knowledge and tools that are the scientific foundation for solving problems in natural resource management.

Essentially all of the major scientific challenges require changes in the way researchers organize themselves as well as improvements in technology, such as remote sensing. It is essential that ecosystem research teams be holistic and interdisciplinary. While scientists working individually or as small groups can contribute to solutions, the major problems, such as global change or even the development of alternative silvicultural systems, require the perspectives and contributions of many disciplines.

Ecosystem ecologists and their colleagues have developed and tested hypotheses about forests on the basis of a systems view of their structure and function. Their promising preliminary work needs to be augmented by greater inclusion of manipulated ecosystems, such as managed forests, and by greater application of ecosystem principles to the management of forests and other renewable natural resources.

Forest Ecosystem Research

Research on the structure and function of forest ecosystems has provided much of the information that is provoking major changes in our view of forests and how they work and has provided many concepts that are helping to resolve conflicts among resource values, such as timber and wildlife. Much of this research has been carried out by scientists outside of the field of forestry with funding from the National Science Foundation and private foundations.

We need to drastically expand the effort devoted to basic research on both natural and intensively managed forest ecosystems. During the past 20 years, such research has been extremely productive of concepts that have revolutionized perspectives on forests such as (1) the structure and dynamics of the below-ground subsystem, including the complex interactions among trees and microbial and fungal communities and the very high rates of turnover and energy use; (2) the scale and intensity of interactions between forest canopy and atmosphere, including the canopy's major roles as condensing and precipitating surfaces; (3) the numerous and critical linkages between forests and associated streams and rivers, including the roles of riparian zones and streamside forests in aquatic productivity and groundwater pollution control; and (4) the rich array of organisms and processes associated with older natural forests.

Current efforts to develop new forest management systems (see Alternative Silvicultural Systems) are drawing heavily on this small body of basic research. Greatly expanded efforts are needed to extend this knowledge to all important forest ecosystems and to deepen our understanding of subsystems and processes, such as those below ground.

Landscape Ecology

Many current forest management problems involve concerns on large spatial scales and must be viewed, at least in part, at the level of a landscape. The preservation of habitat for wide-ranging species, such as large ungulates (for example, elk) or predators (for example, grizzly bear) was an early problem requiring a larger scale perspective. The northern spotted owl in the western United States and the red-cockaded woodpecker in the eastern United States are more current examples. The cumulative effects on water quality and fish habitat, such as undesirable hydrologic and sediment responses resulting from excessive short-term cutting within a particular drainage area, also require a landscape perspective. Another current and important example is forest fragmentation, as in the division of forest landscapes into small-scale patchworks that are less than optimal for wildlife or particularly susceptible to forest catastrophes, such as wind damage.

The developing area of landscape ecology deals with the significance of forest patch size, patch types, the importance of edge or boundary phenomena, and isolation or connectiveness between forest patches of similar types. Expanded research on forest landscape phenomena is critically needed for developing the theoretical basis of landscape ecology and for strengthening the applications of its current concepts.

Further, analytical tools and models are badly in need of development, including the application of geographic information systems (GIS) and expert system technology to habitat classification systems and other inventory and monitoring tasks. Remotely sensed imagery from satellites will be increasingly used for complex predictions of growth, yield, and ecosystem and global change.

Global Change

Global change is one of the most important emerging scientific challenges facing mankind. Global change includes basic changes in climate associated with increasing concentrations of greenhouse gases and pollutants, reduced concentrations of ozone in the stratosphere, deforestation, soil erosion, and declining water quality. The scientific challenges are immense and include the prediction of the direction and intensity of changes, the assessment of the ecological and social consequences of predicted change, and the identification of appropriate societal responses such as measures to mitigate impacts and adaptations to them. Furthermore, these assessments have to be made at scales from local to global and over very long periods of time.

Issues of global change are numerous: How will global change affect the composition, growth, productivity, and distribution of forest ecosystems? How will climatic change affect the emissions of greenhouse gases and natural hydrocarbons from forest ecosystems? Are there feasible mitigation or adaptation strategies for minimizing the effects of global change on forests in a region? How will global change affect the quantity and quality of water from forested watersheds? Resolving or even understanding these issues in sufficient detail that appropriate forest policy can be implemented should be the central focus of much forestry research.

Biological Diversity

The basic concern about loss of biological diversity is the accelerating and irreplaceable depletion of genes, populations, species, and ecosystems.

Associated with this depletion is the possible disruption of essential ecological processes, the loss of products currently or potentially obtained
from natural resources, and the loss of options for biological and cultural adaptation to an uncertain future. Changes in aesthetic quality are
also of increasing concern to society as the natural environment becomes
progressively more uniform and biologically impoverished.

Issues of preservation of biological diversity include some old ones
(such as how to maintain game and anadromous fish populations), some
new ones (such as what to do about threatened and endangered species),
and some that are generally unappreciated (such as the need to maintain
invertebrate diversity and local populations of organisms). Conservation
biology is the label sometimes applied to this rapidly expanding area of
science that includes both theoretical and empirical research. Although
biological diversity has often been thought of as a "set-aside" issue, it is
increasingly clear to many scientists and managers that biological diversity
cannot be dealt with solely by creating reservations. Scientific attention is
being directed to the role of the entire landscape matrix, including lands on
which commodity production is dominant, in maintaining diversity. Hence,
the relationship between reserved and commodity lands is a subtopic of
increasing interest.

Questions associated with the preservation of biological diversity include the following: What elements of biological diversity in a region are
most at risk and where are they located? How do changes in global climate
and atmospheric chemistry and deposition interact with habitat modification to affect biological diversity? How is the loss of diversity at one level
of the hierarchy (such as that of the gene or the species) either associated
with or compensated by changes in diversity at other levels (such as that of
the ecosystem or the landscape)?

Alternative Silvicultural Systems

Development of silvicultural systems based on sound biological and
ecological principles is a major challenge for applied forest research, which
must be directed toward new treatments of individual stands and landscapes. Silvicultural systems that provide forest products while also allowing
for recreation, more structural diversity, and the production of food will
be useful in many settings. Examples are systems that incorporate coarse,
woody debris, integrate tree and food crops, and create multistructured
stands. Successional and ecosystem concepts can provide the theoretical
support for such systems. Dramatically improved knowledge of landscape
structure and function is also needed if we are to develop ecologically
sound alternative silvicultural systems. Major forest issues that impinge on

new silvicultural systems—including biological diversity, global change, cumulative effects, and the protection of fish and wildlife—ultimately require resolution at the landscape level.

Intensive Management for Wood Production

Large tracts of forest land, particularly those owned by the forest industry, will continue over the foreseeable future to be managed primarily for wood production. Many areas of public land are also especially well suited for wood production, which, if concentrated there, relieves the pressure to manage less suitable lands intensively. Even these selected tracts, however, will experience management changes brought about by societal demands. Research questions revolve around two issues: (1) incorporating capacity to generate multiple products along with wood, and (2) learning how to manage forests for wood production without the use of historically successful forestry tools, particularly clear-cutting, chemicals, and fire. The search is on for biologically sound and economically efficient practices that are acceptable to the public.

A close tie exists between silvicultural complexity and harvesting complexity. When timber is grown under silvicultural systems requiring multiple harvests, multilayer stands, and the leaving of woody debris on site for wildlife, the job of removing marketable trees becomes extremely difficult and costly. Developing harvest systems and machines to do this job efficiently and safely will require the application of research. Similar statements can be made about site preparation, slash abatement, and regeneration.

Another requirement for complex silvicultural systems is detailed site-specific planning among silviculturists, forest engineers, fish and wildlife biologists, and other resource specialists. For example, the hazards of certain forest tree diseases, such as annosus root rot, little leaf disease, and *Phytophthora* root rot of Fraser fir, have been shown to be reduced by site selection studies involving analyses of soil characteristics as well as preplanting sampling of soil for presence of the pathogen. Forestry is suffering from a lack of properly trained personnel under present operating conditions, a problem that will only get worse as the silvicultural systems become more complex.

HUMAN-FOREST INTERACTIONS

The needs of people drive the use and the misuse of forests. Our efforts to understand how people think about and act on forests have been minimal, and yet most controversies and shortages ultimately arise from human activity. The role of forests and forestry in rural development is recognized as important, but the research base is inadequate in both developed and developing countries around the world. Cooperative efforts between natural and social scientists in forestry are few. The opportunity to increase knowledge and solve problems is great if research on human-forest interactions is accelerated and if the social ecology of forests is better understood.

Sociology and Forestry

Forests are constantly changing as dynamic social and biological ecosystems. Social change with regard to human conceptions and cultural definitions of forests appears to be accelerating, placing the forestry professional in a position of interacting with traditional and new constituency groups that are competing, with often divergent and conflicting demands, for use of forest lands. Future generations of foresters and forestry educators will need to better integrate knowledge of behavioral science and social-cultural systems into biological conceptions of forests.

Natural resource sociologists interested in forestry have provided an essential, but incomplete, body of information on forest systems. They have sought to understand the adaptive strategies utilized by people as they harvest forests, live within and adjacent to forests, and enjoy the lands set aside as wilderness and parks. Such scientists have studied behavioral dimensions of human-caused fires, connections between forestry and agriculture, wildland recreation, and associated issues of carrying capacity (including volume and patterns of use) and recreation lifestyle. Underlying all such studies is the recognition that forests are social as well as biological systems and that people are integral parts of the definition and use of the forest ecosystems. Several research areas emerge as essential to discussions of a future research agenda linking social science with the biological sciences of forestry. Each will be briefly discussed, and recommendations for research and instructional programs will be provided.

Community Systems.   The interaction of human social systems and resource management has reemerged as an area of scientific inquiry. Studies of rural communities in transition, technological change in forestry and agriculture, and industrialization of the countryside have provided the impetus to study community structure and social systems tied to biological systems.

Community, however defined, provides forest scientists and forest managers an opportunity to understand the direct linkage between humans and various natural resource systems. Community, in contrast to other forms of sociological inquiry, focuses on social structure—the network of institutions providing order to human affairs. The knowledge of process and structure in sociobiological systems is critical to understanding the persistence and change of these rural institutions and ultimately the cultural fabric of community. Better understanding of the human community associated with forests is required for guiding forest management and providing a perspective for sustainable resource development.

Urbanization of the Forests and Urban Forests. Cities are encroaching upon forest boundaries. Conversely, all cities contain trees and many contain forests. These interactions have several effects. First, citizens are becoming more active and involved in the decision-making process about forest management and forest use, often influencing consideration and implementation of alternative forest plans. Second, human habitation within the forest and at its edge is on the rise. Homes built in forests for year-round living are increasing in number. Expectations for public services such as water and fire protection are altering practices of forest fire management. Urban residents who define forests as backyards or vistas are turning forests into parks, with trails, gardens, and recreation superceding forest harvesting practices. Major questions include, At what rate are forests being urbanized? How and where is it occurring? What constraints are being placed on harvesting rates? How are forest management plans and policies changing to accommodate multiple values of the forest expressed in the urbanization of those forests?

Urban forests—wooded tracts of lands in cities and metropolitan areas—provide habitat for wildlife, scenic outdoor space for people, and economic value to cities. Urban forests are far more extensive than most people realize, covering an estimated 69 million acres (Grey and Deneke, 1986). Such forests are of special significance in our highly urban society. Eight out of ten Americans now live, work, and spend most of their leisure time in and around urban areas. Urban forests are particularly important to those Americans who have limited access to more rural areas, including those who are old, young, disabled, disadvantaged, low in income, and short on time, as well as members of minority groups who fear discrimination in more distant areas. Yet these places are little understood. Urban forests are often a diverse component in an even more complex ecosystem. Characterized mainly by trees, but including other plants, animals, and climatic and soil conditions, these places provide habitat for wildlife and people, clean and cool our air, deflect or absorb noise, produce oxygen, and reduce carbon dioxide emissions.

A broad spectrum of benefits and opportunities provided by urban forests ranges from sitting in the cool shade of an urban park to hiking in the "wild" parts of forest preserves and studying nature in arboretums, conservatories, and zoological gardens. Tree-lined corridors linking larger tracts provide a forest setting for increasingly popular "linear activities" such as walking, jogging, bicycling, and skiing.

Urban forests present special problems that need to be solved if we are to succeed in providing and protecting this essential part of the human environment. Both education and research are needed specifically for urban forests. We need to gather information on individual tree species and other organisms as distinct from stands or forests. These data will answer management needs for maintaining these organisms. More knowledge is needed in selecting the best tree species for each urban forest site and in controlling the growth of trees, especially under power lines and in landscape plantings near buildings. We need to know more about which insects and diseases should be controlled and how best this can be accomplished. Pruning, tree removal, and disposal are major issues. Urban forest management practices need to be acceptable to the public as well as to the managers. We need more research on construction damage and how to reduce it substantially; large numbers of trees are being killed or severely damaged because their roots are injured during construction.

Regional Resource Systems. Forestry policies—like those of agriculture, fisheries, mining, tourism, and the protection of natural resources—can no longer be developed in isolation from other potential uses of resources. Forestry at the expense of agriculture, or agriculture at the expense of fisheries, or any primary resource production process at the expense of resource protection ignores the systemic relationships that exist. Conversely, few parks and preserves are large enough to protect a given species within their bounds. Such enclaves depend on the resource management practices about them to achieve their goals. The twenty-first century forester, farmer, government warden, and park manager will need to recognize the interdependence of each of their forms of natural resource management on the others. To succeed, resource management must be considered in the context of an ecosystem, where resource development, conservation, and protection are considered simultaneously. Competition for resources will give way to cooperative management strategies, where conservation and resource management are linked in sustainable resource systems (Field and Burch, 1988).

Two concepts underlying the thinking within a regional planning process and action plan are ecosystem management principles and planning at the landscape or regional level. Research must further define these concepts and their utility for forest management. In addition, a behavioral science–human ecological research perspective on this topic will enhance the future of forestry practices.

Recreation and Aesthetics. Social science research oriented toward recreation should be continued. However, the form and focus of the research should be shifted. The emphasis has been on wilderness and back-country research at the expense of understanding recreation in the general forest system. Further, the nature of the research has been problem-driven, particularly by conflict among different groups that use forests for recreation. For example, few studies exist of the natural history of a recreation activity, of forest recreation at the urban fringe, and of recreation in agroforestry areas. Finally, most forest sociology research has been more applied than basic and more often appears in nonrefereed publications.

Natural Resource Sociology in an International Context. Natural re-source sociologists are participating in greater numbers in international studies of human-forest interactions, including community studies and in-ternational tourism. The systematic study of agroforestry and social forestry, in particular, is in its infancy. Benefiting from studies in anthropology and rural sociology, forestry, agriculture, and aquaculture will become inte-grated as a mosaic of resource activities at the community level and will say much about conservation strategies in the future.

Extension of Forest Sociology. Extension specialists are a primary con-duit through which scientific knowledge about forestry is shared with clien-tele groups. Currently forestry extension specialists are likely to have been educated in the physical and biological sciences and to have worked with the technical aspects of forestry. Contemporary issues and future problems will embrace resource-dependent communities, social conflict and conflict resolutions over multiple values of forests, urbanization of the forest and urban forests, and public involvement in planning for and making decisions about forests. Within this context, the issues associated with the clientele groups interested in forestry are becoming more diverse. The range of extension specialists must be expanded to embrace broader disciplinary backgrounds and to improve communication. Similarly, social scientists working on forestry problems must apply their research results to problem solving, help extension specialists understand the human factors in forestry, and help develop examples of the benefits and negative impacts of changes in forestry practices.

WOOD AS A RAW MATERIAL

The Need for a Major and Sustained Commitment to Forest Products Research in the United States

Wood is a leading industrial raw material in the United States, ac-counting for about 25 percent of the value of all major industrial materials. On a tonnage basis, it exceeds all other structural materials combined. Similarly, it is the principal source of industrial fiber. The demand for forest products is growing: The global demand for timber products grew by 90 percent in the past three decades (USDA Forest Service, 1982) and is projected to grow another 45 percent by the year 2000 (FAO, 1986). The U.S. demand is projected to increase to 20.8 billion cubic feet by the year 2000 (Haynes and Adams, 1985). These increased demands, if imposed on limited supplies, will result in increased prices of products and will thereby have a particularly devastating effect on the affordability of housing, furniture, paper, and other forest products.

The reasons for the popularity of wood as a versatile industrial material are well known. Wood is far less demanding of energy than other industrial materials, such as steel, aluminum, plastics, brick, and concrete. Solar energy produces this industrial raw material, and a large fraction of the energy required to process it into useful products is provided by using the residues of its own manufacturing operations, thus minimizing requirements for fossil fuels.

The United States grows more wood than it consumes, but it has been a net importer of forest products ever since 1916. As U.S. dependence on foreign wood products increases, the importation of forest products becomes a significant and growing negative contribution to the nation's balance of trade problems.

If the supply of industrial roundwood from forests in the United States declines, the forest products manufacturing sector of the domestic system may have to sustain itself on foreign raw material. An alternative scenario is for the entire system to diminish to match the smaller supply of raw materials. If the United States is to participate significantly in meeting its own requirements for forest products, it must either increase its production of industrial roundwood, create a favorable climate for the manufacture of products from imported roundwood, or both. Other sections of this report address the problems associated with the first alternative. This section deals with the problems associated with maintaining a manufacturing sector that can utilize both domestic and foreign raw material supplies.

At present, the United States is not only failing to supply its own raw materials, but it is also losing ground in manufacturing. For example, the United States has lost the market in machinery used in the manufacture of wood products. In the manufacture of pulp and paper, most of the major process and product breakthroughs over the past 15 years have come from overseas; these include continuous digester pulping, high-yield thermomechanical pulping, oxygen-bleaching, and control of dioxin in effluents. South American countries are now replacing the United States as the world's low-cost producer of pulp and paper products, and other developing countries are entering these markets. If the United States is to reverse the trend toward dependency on foreign suppliers for forest products, its research efforts must be substantially expanded and improved, and the production of competent research scientists in the field of wood science and technology must be increased.

Although the United States is a net importer of wood, it has been successful in exporting some of its products (such as softwood plywood) when substantial investments have been made in research and development. Similar opportunities exist for other products. Historically, the United States has been a leader in research in wood science and technology, but that leadership has been declining from a peak immediately after World War II. Significantly, the Wallenberg Award for breakthrough research in forestry and forest product research has been presented only once to a U.S. researcher.

The reasons for the decline in research in wood science and technology are numerous. Despite the benefits of wood as a material, federal research investments are very modest. For example, in 1982, the funding of materials research by the U.S. Department of Agriculture, which focused mainly on wood and some agricultural crops, was reported to be about $30 million dollars, or only 3 percent of the $1 billion dollars spent on materials research overall. In addition, federal in-house research on wood science and technology is essentially concentrated in a single laboratory—the Forest Products Laboratory in Madison, Wisconsin. Federal monies have had little effect on the development of regionally based forestry research at universities. A modest but promising effort to broaden the base of federally conducted research after World War II through the establishment of a Forest Utilization Service at each experiment station was abandoned.

The number of undergraduate programs in wood science and technology at universities is declining. Currently, 23 universities in the United States offer some type of program in wood science and technology; because of declining enrollments, four of these have practically abandoned undergraduate education. Graduate education and research, however, continue to various extents at all of the 23 institutions. Doctoral programs are offered in 22 of the 23. All institutions face difficult prospects on several fronts. The number of students who are U.S. citizens, for example, is declining to such an extent that a significant portion (usually 50 to 60 percent) of the graduate enrollment is currently composed of foreign students (A. Moslemi, Chairman, Accreditation Committee, Society of Wood Science and Technology, University of Idaho, Moscow, personal communication, 1990).

In addition to problems in the supply of graduate students, research laboratories in the field of wood science and technology are generally in need of modernization. Available funding is almost always inadequate to meet this need.

A number of research areas will contribute appreciably to such topics as the recycling of the substantial amounts of wood and fibers that are currently discarded in landfills. Research on enhancing export opportunities can yield significant results in increased exports from the United States. Biotechnology relating to wood also has substantial potential. Combining wood with nonwood materials to develop fire-resistant, durable building and industrial materials is becoming increasingly practicable. These are only a few examples of the many profitable areas of research that can be addressed by the wood science and technology community. Such research has the potential to contribute to a vibrant forest products industry as it supplies literally thousands of products to U.S. and global markets.

Timber Harvesting Research

Timber harvesting is the economic mainstay of many rural communities in the United States and the lifeblood of forest operations in that it provides the funds necessary to build roads and carry out desirable multiple-use forestry practices. Harvesting is also an essential operation for industries using wood and fiber because it is the source of raw material. Additionally, it is the primary silvicultural tool used to achieve a variety of objectives ranging from reforestation to stocking control, and it is therefore a constructive mechanism for maintaining the forest environment and productivity.

Timber harvesting, along with wood-fiber processing and related manufacturing, also presents tremendous opportunities for economic growth and development. But retaining the current levels of industrial activity, let alone realizing growth, will require improvements in timber harvesting technology. The improvements are necessary because timber harvesting is currently (1) potentially disruptive and devastating to the forest environment and (2) expensive relative to the total cost of the raw material.

Despite its important and useful aspects, current harvesting technology can cause problems. Heavy machines may compact fragile soils and root systems, improperly laid roads may cause landslides, and logging may bring about undesirable visual impacts over the short term. These often unacceptable problems have reduced harvesting activity in some areas. Such reductions will eventually reduce activity in wood and fiber processing and manufacturing industries.

Additionally, the expense of timber harvesting in the United States adds more substantially to the cost of the raw material than it does in other countries, potentially putting our wood-based industries at a global disadvantage. This, along with our current inability to eliminate unacceptable environmental disruption, may leave our industries and rural communities unable to capitalize on economic opportunities.

Current problems in timber harvesting represent three engineering challenges: (1) Negative aspects of timber harvest and silvicultural operations must be reduced or eliminated, (2) efficiency of silvicultural and harvest operations must be increased, and (3) the cost of harvested raw material must be reduced.

Engineering research is essential to meeting the challenges and capitalizing on opportunities in timber harvesting and wood-based material processing and manufacturing. Evidence of the potential long-term benefits of timber harvesting research is seen in the Scandinavian countries, where technology has been developed, through research, to make timber harvesting a highly successful component of their profitable export-oriented wood and wood-fiber industries.

Research to Avoid or Minimize Negative Environmental Impacts During Harvesting. Because harvesting damages the forest, it is essential to design harvesting practices and systems that do not reduce long-term productivity, add unwanted sediment and debris to streams, reduce desirable wildlife habitat, or destroy beautiful forest vistas. In fact, our challenge is to enhance outputs of all valuable forest resources and see that they are sustainable and complementary, not merely to keep the production of one output from destroying another. Our goal should be nothing less than higher timber yields, cleaner water, more fish, more diverse native fauna, and better recreational opportunities. Timber harvesting practices affect all of these outputs and must be carefully managed to provide positive influences. Such a result will require a great deal of research to investigate consequences of various harvest systems operating over varied topography and soils, as well as their application in time and space.

Research to Improve the Efficacy of Silvicultural and Harvest Machinery and Operations. We must learn how to undertake forest harvest and silvicultural operations so that the desired outcome is obtained. Whether through greater mechanization, worker training, better planning, or alternative harvest and transportation systems, it is essential that operations in the forest have results that are precisely those intended by the forest manager or harvest plan designer.

Greater mechanization in the woods can mitigate environmental damage if logging practices designed for site-specific topography and stand conditions are used properly. On the other hand, improper use of heavy equipment will exacerbate the problem. Even with more conventional logging practices, the challenge to research is to provide forest engineers with information needed to weigh economic against environmental impacts. Research investigating mechanized harvesting must especially consider impacts on long-term productivity.

A major challenge to forest managers is harvesting on steep slopes, fragile soils, and other sensitive areas that are off limits to conventional ground equipment. Flexible harvest systems (cable, aerial, and mixtures with ground systems) having low ground impact and capable of performing logging operations over long distances and irregular terrain must be developed and tested.

Research to Reduce Harvesting Costs and Improve Profitability and Safety. Research into productivity improvement and cost reduction is important because the timber supply is changing—often to smaller trees and trees of less desirable shapes and of less desirable species. These changing attributes are often associated with higher logging costs and, in turn, with businesses having greater difficulty remaining profitable and competitive.

People are the most important part of any organization, and the logging industry faces an especially serious safety and training problem. Workers' compensation bills, which can approach the value of gross payroll, dictate a strong need for greater emphasis on training and safety. New logging machinery requires sophisticated operator skill. Major improvements in training and safety in the logging industry are possible over the next decade.

INTERNATIONAL TRADE, COMPETITION, AND COOPERATION

Improvements in international trade, competitiveness, and cooperation in the sustainable production of goods and services from forest resources depend on information. Detailed knowledge about forests and natural resources worldwide is limited at best. Information about international trade is also sketchy, yet trade policies that facilitate economic growth while sustaining the natural resource base are urgently needed. Forest resource inventory and commercial supply characteristics of participants in world markets must be made available if meaningful work on integrated worldwide supply and demand projections is to be initiated.

Information, Supply, and Demand

Changes in international trade policies, national marketing strategies, and world markets can have significant impacts on the natural resource base. Economic models must be developed that can predict and assess these changes and their consequences on the domestic as well as the worldwide natural resource base. The role of government policies on domestic markets, international trade, competition, and cooperation in production and distribution of forest outputs of goods and services must be explored.

Economic models must, over time, describe and analyze forest resource ownership by such characteristics as size, distribution, and managerial objectives. An important goal here is greater understanding of the potential for production of nonmarket goods and services from forest environments and increased ability to assess the value of these products. A corollary objective is the exploration of domestic and international policies that encourage trade in nonconsumptive uses of forest resources.

An opportunity exists to obtain matching support for this type of economic analysis. This is possible through existing legislation for the creation of International Trade Development Centers (ITDCs). This legislation requires interdisciplinary approaches and collaboration between state and federal agencies. Although many states have established promotion and contact programs, most find themselves short of analytical underpinning; implementation of ITDCs, particularly on a regional basis, could help to fill that void.

Substantial effort is needed to explore the impacts of environmental constraints on the production and export of wood products. Such constraints can raise costs. Increases in prices of wood products force greater substitution of commodities that depend on nonrenewable natural resources. Currently, the United States can increase supplies of forest products with less environmental degradation than most other nations. Environmental restrictions imposed worldwide could cause developing countries to depend increasingly on U.S. forest products. Environmental constraints imposed on U.S. production, however, could cause increased exploitation elsewhere with detrimental environmental consequences. Research is needed to explore how international trade and debt policies can foster equity among nations in both benefits and costs of environmentally sound management of natural resources coupled with sustainable economic growth.

Basic to trade in any market is the value of items exchanged. We clearly need to increase our understanding of prices and values in the exchange of goods and services produced from forests. Many wood products reflect prices that are consistent with their true value; many do not. Noncommodity products of forests are often not priced in accordance with their value, nor are they appropriately recorded in national accounting systems. In-depth studies are needed to understand the role and value of forests in broader environmental issues such as water quality, the carbon cycle, ozone levels, and air pollution. Follow-up studies would provide for the proper accounting of both market and nonmarket values and associated costs of outputs of forest environments.

Utilization, Marketing, Employment, and Exchange. Significant gains can be captured from domestic and international wood utilization research to make more efficient use of existing wood supplies. More effective use of forest resources, especially hardwoods, would extend supplies, reduce waste, and slow deforestation. In a number of situations, technological processes in the temperate climates of the United States and other industrial nations may be adapted readily to local conditions and species in the tropics. Returns on investments in enhanced utilization of wood supplies can be high.

Forest-based industry is critical to employment, markets, and international exchange. Forest industry can contribute to local and national economies in several ways. Employment and income are generated, inputs for other sectors of the economy are provided, and foreign exchange is augmented. These contributions, however, depend on basic forest resources, depletion of which can seriously threaten the supply of both wood and nontimber goods and services. Internationally, this threat is real. Tropical deforestation and over-exploitation of natural forests, wasteful harvesting practices and inefficient utilization, and inadequate investment in forest management and reforestation call for both research and action to manage and reduce threats to the resource base.

Increased imports impose a high cost on developing nations. Many of these countries have a sufficient base in land and natural resources to meet their domestic industrial and consumption needs at a cost lower than that of import alternatives. These countries and the international community would benefit from major research efforts aimed at enhanced management of forests and promotion of appropriate and sustainable forest-based industries. Major components of this research would be on natural regeneration and perpetuation of species having high commercial value and on improved harvesting systems to increase utilization and reduce logging damage.

International trade, competition, and cooperation depend on prices and accounting systems that accurately reflect the value of the forest resource and its products. A major need exists for research on appropriate and valid concession and pricing policies, especially in tropical countries. Existing research demonstrates that some government policies generate strong economic incentives to accelerate deforestation. Identified research needs include (1) understanding the level and structure of timber royalties and other charges; (2) exploring forest concession policies, including their duration, other license fees, and prescriptions of harvesting methods; and (3) examining trade-offs between policies that encourage export of logs versus those that facilitate domestic processing.

Interdependence and Externalities. The high interdependence of all natural resource systems makes production and distribution decisions extremely difficult. Management decisions impose costs and benefits not only among owners, neighbors, and world markets, but between present and future generations of humankind. These questions and implications of interdependence and externalities call for research that integrates a number of disciplines. The stakes here are substantial. They include the quality of air, water, and soil resources, the biological diversity of natural and managed systems, and the sustained productivity of the world's forests, crop lands, grazing lands, wetlands, and riparian zones.

International Competition and Cooperation. Population growth and increased worldwide poverty call for coordinated and cooperative remedial efforts. Population pressure and poverty often result in accelerated deforestation, especially in developing countries. Critical research questions about the impact of growing populations and poverty on international trade in forest products must be investigated in terms of supply and demand projections and prospects. An underlying issue here is the question of whether or not the world's natural resource base can both support population growth and contribute to the diminution of poverty. With an objective of sustainable economic development worldwide, research must identify nations or resource conditions that promise intrinsic comparative advantage in wood and forest products.

The United States has traditionally played a major role in exports and imports of forest products. We need research that would clarify the role of currency exchange in determining international trade. When the U.S. dollar falls in value, U.S. products become relatively inexpensive and exports tend to increase. At the same time, the underlying economic resources, forest land, and production facilities become more valuable as assets to be purchased by foreign investors. Research can help to identify the implications of such prospects. Furthermore, studies can be directed at such questions as why the United States exports logs and imports finished products. Answers will offer strategic insight into actions that trading partners can take to ensure continued gains from increased competition and cooperation.

Policy Research. Better international trade, competition, and cooperation can be ensured only with resource policies that facilitate attainment of these goals. Both national and international policies for resource use and management are relevant. Policies must be preceded by research that examines some critical questions: (1) What are the effects of major macroeconomic entities such as the Federal Reserve on forest resources? (2) Why are forest resources in developing countries being depleted so rapidly? (Is it national debt? Agricultural policies?) (3) Reduced deforestation in tropical countries implies economic advantage for industrialized nations in the forestry sector—how can this conflict of interests be rectified? (4) What are the forest resource implications of a global carbon tax based on net national emissions of carbon? Forestry research directed toward major societal issues can contribute significantly to progress in formulating policies that ensure sustainable development.

4

Conclusions and Recommendations

The preceding chapters report some of the diverse activities and needs of forestry research. We have described research on wood as a raw material and in basic biology, ecology, sociology, and economics and policy. Although forestry research is at least as complex as agricultural research, to which it is closely related, forestry research cannot be subsumed under what has been traditionally viewed as agricultural research, but must be viewed as having values for society that are broader than and distinct from those of traditional agriculture (National Task Force on Basic Research in Forestry and Renewable Natural Resources, 1983). In this chapter, we provide several broad conclusions and attendant recommendations concerning the nature of forestry research, human resources, ways of maximizing the benefits from forestry research, and support for forestry research.

THE NATURE OF FORESTRY RESEARCH

More Scientists Should Do Forestry Research

Conclusion. For forestry research to be of sufficient quality and quantity to solve critical societal problems, it must embrace more areas of science. These include not only such expected areas as biology, hydrology, and engineering, but also economics and sociology.

Conclusion. Forestry research is overly fragmented by disciplines that interact insufficiently. Interaction must not only increase among traditional forestry disciplines within colleges of forestry, but also among disciplines within colleges of agriculture and colleges of arts and sciences. This need for increased disciplinary interaction coupled with increases in the cost of research facilities, the specialization of scientists, and the diversity of sponsors and clients argues for aggregation and integration of forestry research.

Conclusion. With numerous advisory committees representing organizational research interests, leadership in forestry research has been fragmented. Government agencies and other organizations responsible for research activities can obtain policy advice from a wide variety of sources, such as internal advisory committees at various levels within a department's hierarchy. Research organizations can also draw upon other groups, such as the National Research Council, for advice. Because of the broad range of research organizations and clientele of forestry research, none of the existing forestry advisory committees has adequately met the needs of the forestry research community in general.

Recommendations:

• Provide a vastly expanded funding mechanism, such as competitive grants, to support scientists now doing forestry research and to attract additional ones.

• Strengthen and broaden the teaching of forestry to attract a broader array of students, especially at the graduate level, and to interest other on-campus research groups.

• Establish a National Forestry Research Council (NFRC) to provide a forum for deliberations on forestry research and policy issues. The NFRC should be convened under the auspices of an organization or organizations that can facilitate discussion and action. Financial support for the council's activities should come from member organizations and other interested sponsors. The NFRC should consist of representatives from major organizations—such as government agencies, industry, conservation organizations, private foundations, and academia—with strong interests in forests and related renewable natural resources and in agriculture. The NFRC would commission studies, conduct analyses, and provide advice to policymakers on issues pertaining to those interests.

• Encourage conservation groups and other nongovernmental organizations to more actively support basic forestry teaching and research through the activities of the proposed NFRC.

Conclusion. Coordination and integration with other research scientists should be increased. The proposed NFRC could provide leadership and a forum for coordination and integration. In addition, integration can be achieved by creating centers of emphasis on research in the five areas of research discussed in this report (biology of forest organisms; ecosystem function and management; human-forest interactions; wood as a raw material; and international trade, competition, and cooperation). The creation of a center of emphasis does not necessarily require the construction of a new research facility. It does require, however, a cooperative mechanism for research that allows scientists to interact in a manner that enhances their productivity. In addition, Forest Service scientists could be routinely placed within university academic units, such as departments of botany and zoology, as well as within schools of forestry. A successful model for this type of integration is that commonly used by the USDA Agricultural Research Service.

Recommendation:

• Create centers of scientific emphasis involving major participants in forestry research for each of the five research areas discussed in this report. More than one center could be established for each of the five research areas, depending on the particular interests and strengths of the proposed center's participants.

Result. Benefits derived from a broader definition of forest research and the inclusion of nontraditional research pursuits will include increased relevance to society as a whole, higher intellectual achievement resulting from broader spheres of influence, enhanced attractiveness of the profession to talented scientists, and increased political support for research programs.

The consequences of failing to incorporate these pursuits and personnel into the forest research establishment will be great: They will include not only the decreased quality of forestry science and technology, but also further erosion of public confidence in the relevance of forestry to society.

Adopt a New Approach to Forestry Research

Conclusion. To help overcome a deficiency in knowledge, a new research paradigm will need to be adopted—an environmental paradigm. Past approaches to forestry research employing the conservation and preservation paradigms have proven inadequate.

Conclusion. Many issues—such as biological diversity, cumulative effects of pollutants, and land use or land management—must be addressed at very large spatial scales and over long periods of time. Research at the scale of landscapes and regions will involve changes in the way forestry research is performed, including new mapping technology, collaboration with managers and user groups, creative experimental approaches, and evaluation procedures that differ dramatically from those of traditional forestry research.

Conclusion. Most of the research needs highlighted in this report are as relevant to tropical as they are to temperate forestry. Research related to deforestation and loss of biological diversity are especially relevant to the tropics, but forestry research should derive principles that apply across bioclimatic zones.

Recommendations:

• Establish research-management collaborations at large spatial scales with an environmental perspective. This will require multidisciplinary activities on large tracts of land.

• Establish long-term forestry research (LTFR) grants to provide a peer-reviewed, competitive funding mechanism for long-term research support (longer than one forest rotation).

Result. Scientists and managers will collaborate to develop, install, test, and revise practices on large blocks of land, each block unique in its set of environmental and social conditions.

HUMAN RESOURCES

Conclusion. A critical need exists for the forestry research and policy community to open its ranks to participation by scientists who are often not now considered forest scientists. Contemporary issues, such as sustainable development, the role of forests in global carbon balance and global warming, acid rain, and the preservation of biological diversity, illustrate the need for scientific expertise inadequately represented by traditional forest science.

Conclusion. To meet future demands for research and education, a large number of well-educated scientists, technicians, extension specialists, and educators are needed. To help meet the future need for talented scientists, significantly more women and members of minority groups must be recruited into forestry research.

Conclusion. Forestry education should be restructured to place more emphasis on fundamental research tools.

Recommendations:

• Enhance the quality of forestry research by opening it to the broader scientific community and encourage increased participation by scientists currently within the community.

• Establish a program to provide doctoral fellowships on a competitive basis for all areas of forest and environmental sciences. The program should be designed to attract the highest caliber of students possible and to provide numbers of scientists with appropriate skills to meet impending needs. This program should be supported at a rate of $5 million per year, which will support a total of 200 doctoral fellows per year for 5 years (40 new fellowships awarded each year).

• Develop a cadre of forest and related scientists that reflect the national and global population composition and that are equipped to solve domestic, international, and global problems. A recruitment program for women and members of minority groups should be directed toward high school, undergraduate, and graduate levels and should provide internships and fellowships (as part of the doctoral fellowship program).

Result. With the addition of scientists from broader cultural and scientific backgrounds, forestry research will become more relevant to the needs of society, more interactive with other scientific disciplines, and more productive in developing the needed base of information for making better decisions on natural resource policies.

MAXIMIZE THE BENEFITS FROM INCREASED FORESTRY RESEARCH

Conclusion. The demand for scientifically based information and expert opinion on environmental issues and human-forest interactions will continue to increase. Forest scientists are responsible for keeping the public informed about the status of forests and global environmental issues. Forest scientists need (1) to improve their communication of research results to the public and natural resource professionals and (2) to increase their assistance and involvement in the formulation of policy.

Recommendations:

• Incorporate an outreach component into research projects to communicate results to a broader range of clients.

• Establish a professional reward system to acknowledge the validity of efforts of scientists involved in outreach.

• Scientists should assume a leadership role in communicating their knowledge to those involved in policy-making.

Conclusion. A strengthened program in forestry research requires a greatly strengthened and reorganized companion extension outreach effort. The broadening of program directions into areas as diverse as urban forestry and molecular biology will require additional support and a larger and more diverse cadre of extension specialists capable of communicating ideas as well as techniques. Extension forestry is an important mechanism for technology transfer and education, particularly to nonindustrial forest landowners, natural resource professionals, policymakers, city planners, and the public. The present extension forestry infrastructure is inadequate to serve current or future needs. For example, the Renewable Resources Extension Act (RREA), passed in 1978, has been funded at only about 20 percent of its authorization level.

Recommendations:

• Double the base level of funding and number of full-time equivalents (FTEs) devoted to forestry extension in cooperation with state and local partners.

• Increase RREA funding to the appropriation authorization level of $15 million dollars annually.

• Integrate extension specialists with their research counterparts at colleges and universities in those instances where interaction between extension specialists and research scientists is inadequate.

Result. With adequate knowledge and technology transfer mechanisms, the results of forestry research can inform and instruct a broader clientele, including natural resource professionals, youth, policymakers, urban dwellers, conservation organizations, and the public.

SUPPORT FOR FORESTRY RESEARCH

The recommendations contained in this section are based on the committee's own study and knowledge of the U.S. forestry research system, on interviews with additional scientists, and on documents the committee received from forestry-associated research organizations. The funding increases recommended in this report reflect the committee's experience in and concern about the current status and future prospects of forestry research in the United States.

Equipment and Facilities

Conclusion. The committee believes that the physical facilities and research equipment at many forestry research stations and forestry colleges are inadequate. Other reports assessing the status of equipment in biology (NIH, 1985) and agriculture (Biggs *et al.*, 1989) have drawn similar conclusions. Laboratories lack essential resources to carry out state-of-the-art research in the forest sciences. For example, facilities and equipment needed include electron and video-enhanced microscopes, computers, geographic information systems, greenhouses, and plant-growth facilities. Funding has been inadequate to keep pace with changing technology; therefore research and teaching are not up to date.

Recommendation:

• Conduct a national assessment of the current status of equipment and facilities needed to carry out the research described in this report.

Funding

Conclusion.   Recommendations for increases in funding for forestry research come at a time of overall fiscal constraint for the nation. Government officials must both reduce the national debt and set priorities among competing federal expenditures to enact programs that maintain the welfare, infrastructure, security, and continued economic growth of the United States. As a part of that endeavor, they must also address public concerns for maintaining global competitiveness and environmental resources. The goal of reducing expenditures while allocating funds for essential programs thus requires fiscal prudence.

The committee recognizes that current federal budgetary constraints make new funds for research support exceedingly difficult to obtain. Meaningful increases in research support for forestry and forestry-related research will likely be realized only as a result of changes in funding priorities within the U.S. Department of Agriculture (USDA) and the U.S. Department of the Interior. As outlined in this report, the need to make these changes in funding priorities is urgent if future forests and related renewable natural resources are to be protected from misuse and environmental degradation and if productivity is to be enhanced.

Conclusion.   The largest centrally administered forestry research budget is that of the USDA Forest Service. Therefore, if forestry research is to be reshaped and augmented as described above, changes in this budget are imperative. Additional changes in other forest research funding mechanisms, such as McIntire-Stennis and USDA competitive grants, are also imperative. Funds available through such programs as McIntire-Stennis should be used in creative new ways and to a greater extent to attract relevant scientists from outside forestry schools and colleges. Competitive grants should allow for research flexibility to fund both short-term (2 to 5 years) and long-term (7 to 10 years, or longer than one forest rotation) research programs.

Conclusion. Both the Forest Service and the Cooperative State Research Service of the USDA need to compensate for losses in research support caused by budget cuts and inflation and should play leading roles in establishing centers of emphasis. Industry, state, and private sources of support should also contribute to this effort.

Recommendations:

• Increase USDA competitive grants for the five major research areas discussed in this report with a provision for LTFR grants. To cover the five areas (the biology of forest organisms; ecosystem function and management; human-forest interactions; wood as a raw material; and international trade, competition, and cooperation), approximately $100 million annually will be necessary. A logical basis for this type of competitive financial support is through the current research funding initiative proposed by the NRC's Board on Agriculture (NRC, 1989c). The Board on Agriculture report defines agriculture to include forestry and related areas. As proposed, this initiative identifies natural resources and the environment as one of six program areas that need increased funding. Four other identified program areas (plant systems; animal systems; engineering, products, and processes; and trade, marketing, and policy) are directly related to the forestry research described in this report. The total amount requested in the Board on Agriculture research initiative for USDA competitive grants is $500 million annually. Traditionally, however, forestry research has not been granted proper status in the USDA competitive grants programs. Therefore, for forestry and forestry-related research to be adequately supported by the results of the Board on Agriculture research initiative, changes in funding philosophy must take place within the USDA.

• Increase the USDA Forest Service research budget by 10 percent each year for the next five years. These new funds should be allocated among the five research program areas discussed in the report. With these five successive annual increments, the Forest Service research budget will expand from its 1988 level of $135 million to $218 million after five years.

• Increase McIntire-Stennis funds over the next five years to the full authorization level of 50 percent of the Forest Service budget. These new funds should also be allocated among the same five research program areas discussed in the present report. With these five successive annual increments, McIntire-Stennis funding will expand from its 1988 level of $17.5 million to $109 million after five years.

Result. These three recommendations for increased federal support for forestry research will provide for orderly growth from the present $187 million annually to a total after five years of $427 million annually. After five years, this will mean that annual investments in forestry research will have reached about 20 percent of the total of about $2.5 billion for all agricultural research, after addition of the $500 million competitive grants program recommended in the Board on Agriculture funding initiative (NRC, 1989c).

If these modifications in the forestry research funding are made, forest scientists will be able to provide better advice to the American public on the management of our nation's forests; industry will have a far greater data base from which to improve wood production practices and new forest products; and society in general will benefit from improved global environmental management.

SUMMARY

Forestry research must change radically if it is to help meet national and global needs. It must become broader in its clients, participants, and the problems it examines, and at the same time it must conduct more in-depth research and become more rigorous in utilizing all of science and technology. The number of scientists and amount of resources devoted to forestry research are declining, even as needs increase. To meet the challenge of rapid change, new approaches and new resources of the kind described in this report are required. The educational and fiscal systems that support forestry research must be restructured and revitalized; integrated research facilities must be created where public and private resources can be concentrated on basic questions, new technologies, and effective outreach and extension activities. These changes will be expensive, difficult, and painful for many. They will be painful in that research resources will need to be redirected and certain research facilities may need to be closed. The consequence of failing to make the changes, however, would be even more painful: a national and global society increasingly unable to preserve and manage forest resources for its own benefit and for the benefit of future generations.

We emphasize here that both the misuse and the wise use of forests are consequences of human activity. In the absence of policy alternatives provided by a large increment of knowledge resulting from forestry research, the misuse exemplified by deforestation, destroyed productive potential, and lost biological diversity will prevail. Knowledge gained from an improved system of forestry research will enable society to choose wise use and thus to secure the environmental, economic, and spiritual benefits of forests.

References

AFC (American Forest Council). 1987. Forest Industry–Sponsored Research Cooperatives at U.S. Forestry Schools. Washington, D.C.: AFC.

Backiel, A. 1986. The Renewable Resources Extension Act. Washington, D.C.: Congressional Research Service, Library of Congress.

Biggs, R. H., L. W. Moore, and R. A. Dreher. 1989. Instrumentation and Equipment Survey for Agricultural Biotechnology. Committee on Biotechnology, Division of Agriculture. Washington, D.C.: National Association of State Universities and Land Grant Colleges.

Chapman, R. L., and J. G. Milliken. 1988. Forest Service Research: Dealing with the Issues Underlying Concerns of Competitiveness and System Responsiveness. Fort Collins, Colo.: U.S. Forest Service, Rocky Mountain Experiment Station.

Commoner, B. 1971. The Closing Circle. New York: Knopf.

CSRS (Cooperative State Research Service), Office of Grants and Program Systems. 1989. Food and Agriculture Competitively Awarded Research and Education Grants—Fiscal Year 1988. Washington, D.C.: USDA. (Available from CSRS Information Office, Room 328, 901 D Street, S.W., Washington, D.C. 20250).

Culhane, P. J. 1981. Public Lands Politics. Baltimore: Johns Hopkins University Press.

FAO (Food and Agriculture Organization of the United Nations). 1986. Forest Products, World Projects. Rome: FAO.

Field, D. R., and W. R. Burch, Jr. 1988. Rural Sociology and the Environment. Westport, Conn.: Greenwood.

Forestry Education Problem-Assessment Steering Committee. 1987. Status Report: Problem Assessment of Professional Forestry Education in the U.S. (Supported by the USDA Office of Higher Education and by the University of Minnesota College of Forestry.)

Foundation Center. 1989. Grants for Environmental Law, Protection, and Education. Washington, D.C.: The Foundation Center.

Giese, R. 1988. Forestry research: an imperiled system. J. For. 86(6):15–22.

Grey, G. W., and F. J. Deneke. 1986. Urban Forestry. New York: Wiley.

Haynes, R. W., and D. M. Adams. 1985. Simulation of the Effects of Alternative Assumptions on Demand-Supply Determinants on the Timber Situation in the United States. Washington, D.C.: USDA Forest Service, Forest Resource Economic Research.

Hays, S. P. 1959. Conservation and the Gospel of Efficiency. Cambridge, Mass.: Harvard University Press.

Leopold, A. 1949. The land ethic. Pp. 201–226 in A Sand County Almanac and Sketches Here and There. New York: Oxford University Press.

Marsh, G. P. 1864. Man and Nature. 1965 edition, David Lowenthal, ed. Cambridge, Mass.: Belknap.

Mergen, F., R. E. Evenson, M. A. Judd, and J. Putnam. 1988. Forestry Research: A Provisional Global Inventory. Yale University Economic Growth Center, New Haven, Conn. Chicago: University of Chicago Press.

NIH (National Institutes of Health). 1985. Academic Research Equipment and Equipment Needs in the Biological and Medical Sciences. Bethesda, Md.: NIH.

NPCA (National Parks and Conservation Association). 1989. National Parks: From Vignettes to a Global View. Washington, D.C.: NPCA.

NRC (National Research Council). 1986. Pesticides and Groundwater Quality: Issues and Problems in Four States. Washington, D.C.: National Academy Press.

NRC (National Research Council). 1988. Toward an Understanding of Global Change: Initial Priorities for the U.S. Contributions to the International Geosphere-Biosphere Program. Washington, D.C.: National Academy Press.

NRC (National Research Council). 1989a. Alternative Agriculture. Washington, D.C.: National Academy Press.

NRC (National Research Council). 1989b. Field Testing Genetically Modified Organisms: Framework for Decisions. Washington, D.C.: National Academy Press.

NRC (National Research Council). 1989c. Investing in Research. Washington, D.C.: National Academy Press.

NRC (National Research Council). 1989d. Summary Report 1988: Doctorate Recipients From United States Universities. Washington, D.C.: National Academy Press.

National Task Force on Basic Research in Forestry and Renewable Natural Resources. 1983. Our Natural Resources: Basic Research Needs in Forestry and Renewable Natural Resources. Moscow, Idaho: Forest, Wildlife, and Range Experiment Station, University of Idaho.

Potter, V. R. 1988. Global Bioethics: Building on the Leopold Legacy. East Lansing, Mich.: Michigan State University Press.

Rogers, D. L., E. T. Bartlett, A. A. Dyer, and J. M. Hughes. 1988. Research and Education Systems for Renewable Resources, vol. 5. Fort Collins, Colo.: Colorado State University.

Science and Education Administration, U.S. Department of Agriculture. 1980. A Five-Year National Plan for Renewable Resource Extension Programs. Washington, D.C.: U.S. Government Printing Office.

Society of American Foresters. 1988. 1988 Forestry School Enrollment and Degrees Granted Survey Tables. Bethesda, Md.: Society of American Foresters.

Tien, M., and T. K. Kirk. 1983. Lignin-degrading enzyme from the Hymenomycete Phanerochaete chrysosporium Burds. Science 221:661–663.

USDA (U.S. Department of Agriculture, Extension Service). 1986. Renewable Resources Extension Program: Five-Year Plan, 1986–90. Washington, D.C.: U.S. Government Printing Office.

USDA (U.S. Department of Agriculture, Cooperative State Research Service). 1987. Cooperative Forestry Research Advisory Council: 1987 Annual Report. Washington D.C.: USDA.

USDA (U.S. Department of Agriculture, Forest Service). 1982. Analysis of the U.S. Timber Situation, 1950–2030. Washington D.C.: U.S. Government Printing Office.

USDA (U.S. Department of Agriculture, Forest Service). 1989. Budgetary Book of Notes. Washington, D.C.: USDA.

U.S. Department of Commerce, Bureau of the Census. 1989. National Data Book and Guide to Sources: Statistical Abstracts of the United States 1989, 109th ed. Washington, D.C.: U.S. Government Printing Office.

Wilson, E. O., ed. 1988. Biodiversity. Washington, D.C.: National Academy Press.

World Commission on Environment and Development. 1987. Our Common Future. New York: Oxford University Press.

A

Workshop on "The Future of Forestry Research"

April 17, 1989
NATIONAL ACADEMY OF SCIENCES
Washington, D.C.

PARTICIPANTS

Darius Adams, University of Washington
Scott Berg, American Forest Council
Kent Blumenthal, National Recreational Parks Association
Douglas Crutchfield, Westvaco
Gregory Dilworth, U.S. Department of Energy
Barry Flamm, Wilderness Society
Donald Gilmore, Technical Association of Paper and Pulp Industry
Yaffa Grossman, Ecological Society of America
Dwight Hair, American Forestry Association
Tom Hamilton, USDA Forest Service
Charles Harden, Society of American Foresters
John Heissenbuttel, American Forest Council
Fred Kaiser, USDA Forest Service
Stan Krugman, USDA Forest Service
George Leonard, USDA Forest Service
Arnett Mace, National Association of Professional Forestry Schools and
    Colleges
Brian Payne, USDA Forest Service
Boyd Post, USDA Cooperative State Research Service
Paul Ringold, U.S. Environmental Protection Agency
Chris Saint, U.S. Environmental Protection Agency

Roger Sedjo, Resources for the Future
Richard Skok, University of Minnesota
Richard Smythe, USDA Forest Service
John Spears, World Bank
Jack Sullivan, U.S. Agency for International Development
Larry Tombaugh, National Association of Professional Forestry Schools
    and Colleges
Harold Walt, California Board of Forestry
Barbara Weber, USDA Forest Service
Henry Webster, National Association of State Foresters

B

Research Needs

BIOLOGY OF FOREST ORGANISMS

Physiological and Genetic Bases of the Mechanisms Underlying Forest Health, Productivity, and Adaptability

- To determine the physiological and genetic mechanisms involved in the growth and development of important tree species and other key forest organisms, especially their responses to stress and changing environments.
- To determine the ecological functions of all forest organisms that are critical to the maintenance of forest ecosystems and sustainable forestry.
- To determine the factors that limit population viability of aquatic and terrestrial organisms.
- To understand the genetic basis for the unique biology of forest organisms, many of which are perennial woody plants or microorganisms that cannot yet be cultured in the laboratory.
- To develop new model systems that allow forest organisms to be investigated in the laboratory or in controlled environments to facilitate studies of forest biology and forest ecosystems.
- To define genetic and environmental factors that determine the characteristics of a forest environment, particularly nutrient uptake, water uptake, and the effects of these on the physiology of forest organisms.
- To use molecular techniques to modify or select improved organisms that will be increasingly specialized for specific uses. Examples include trees modified for faster growth or improved wood properties.

- To understand the genomic structure of key forest organisms to facilitate identification of valuable genes or to learn how genetic modification could improve specific characteristics.
- To understand the genetic structure of the populations of important forest organisms and to determine the optimal strategy for conservation or utilization of the forest resource on the basis of the potential effects on the genetic structure of the population.
- To utilize the technology of molecular biology to understand the basis for the properties and behavior of cells and tissues in culture.
- To develop molecular markers that can monitor environmental stresses in forest organisms and provide early information for evaluating potential threats to the health of a forest ecosystem.
- To examine the genetics of adaptability and productivity of currently and potentially important tree species or hybrids.
- To identify key organisms (including insects and microorganisms—for example, mycorrhizal fungi) that play important roles in forest ecosystems and to obtain basic information regarding the genetics and physiology of these organisms and the underlying mechanisms by which they interact with and contribute to other important forest components.
- To obtain basic genetic and physiological information on key organisms, including tree pathogens, that stress plants, with a view to either managing the populations of those organisms or reducing their pathogenic effects.

Long-Term Site Productivity

- To understand the factors that regulate site productivity.
- To determine the effects of even-aged management and rotation age on long-term site productivity.
- To determine the effects of various harvesting and intensive management practices on long-term site productivity.
- To increase information on fundamental forest biology, carbon assimilation and allocation, water uptake and nutrient relations, and disease resistance relevant to maintaining and increasing forest productivity.

Pest Management

- To understand more fully the basic biology and ecology of forest pest organisms.
- To determine modes of action for host-plant resistance.
- To establish methods of controlling pests while minimizing the development of resistant pest biotypes.
- To establish risk-assessment data for the introduction of genetically engineered biological control agents.

ECOSYSTEM FUNCTION AND MANAGEMENT

Forest Ecosystem Research

• To expand knowledge of the structure and dynamics of below-ground portions of forest systems, including the roles played by fungi and other soil organisms in controlling productivity and nutrient cycling.

• To expand knowledge of the structure and dynamics of forest canopies, including their roles as habitat for other organisms and as condensing and precipitating surfaces for atmospheric materials, including water and particulates.

• To expand knowledge of long-term changes in the composition, structure, and function of forest ecosystems as they are associated with natural succession and with silvicultural manipulations.

• To examine relationships between forest structure and types and levels of ecological processes (such as productivity) and of organismal diversity, including greatly expanded consideration of organisms other than higher plants and vertebrates.

• To compare the structure and function of natural forest ecosystems and their long-term responses to natural catastrophes and disturbances created by human activities.

• To develop predictive models (including habitat classification) that allow comparison on a range of scales from fine resolution (tree or stand) to forest, biome, and interbiome.

• To develop methods for early diagnosis of stressed ecosystems.

Landscape Ecology

• To develop models and other analytic tools for assessing effects of various landscape patterns on resource values over long periods of time.

• To expand information on the effects of forest patch sizes and configurations on such forest attributes as biological diversity. Special attention should be given to the extent and importance of edge or boundary effects on organismal diversity and on catastrophic forest disturbance.

• To develop information and analytic tools for analyzing cumulative effects of forest harvests on hydrologic phenomena and wildlife habitats.

• To expand information on the importance of connectiveness in forest and riverine landscapes, including the use of both corridors and altered matrices in the movement of organisms and materials.

Global Change

- To analyze the effects of the extent, magnitude, rate, and timing of global change on different forest ecosystems and to identify which forest ecosystems are at greatest risk from changes in climate, ozone, acid rain, and deforestation.
- To determine how the effects of global change interact with management practices, such as intensively managed timberlands, parks, watersheds, and timber production.
- To develop methods to proportionally allocate forest stresses to natural and anthropogenic causes.
- To determine how global change is affecting the release of carbon from forest vegetation and soils.
- To determine stress-induced changes in natural emissions of volatile organic compounds by vegetation.
- To investigate the ecological, economic, social, and operational constraints on large-scale reforestation and other silvicultural practices that could be implemented as mitigation or adaptation strategies for global change.
- To determine the effects of global change on the natural disturbances (fire, wind, drought, and so forth) experienced by different regional forests.
- To analyze the effects of global change on the amount, timing, and chemistry of water yields from forested watersheds.
- To establish sites along climatic and pollution gradients at which to monitor and evaluate environmental stress and to study manipulated ecosystems over the long term.

Biological Diversity

- To analyze interactions between management practices and biological diversity across levels, from the genetic to species and regional systems.
- To determine the relationship between biological diversity and forest health.
- To document and monitor the status and distribution of species, vegetation types, and other levels of biological diversity within forest regions.
- To define the relationship between components of biological diversity (for example, species richness and threatened and endangered species) and species distribution in relation to changing climatic conditions.
- To define the importance and ecological significance of native, coevolved elements of biological diversity versus those that are alien.

Alternative Silvicultural Systems

• To develop silvicultural and agroforestry systems that provide for integrated production of commodities and for maintenance of ecological values.

• To develop silvicultural systems for management of forests for maintenance or enhancement of biological diversity.

• To develop silvicultural systems that will ensure a high level of sustainable productivity by conserving key site resources, including physical and nutritional properties of soil.

• To investigate the relationship between forest structure and the silvicultural systems that provide for greater degrees of structural diversity and ecological values.

• To determine the implications of alternative silvicultural systems for disease, insect, animal damage, brush invasion, wildfire, and logging safety.

• To examine silvicultural patterns, including size of areas selected for treatment and contrasting effects of dispersing versus aggregating harvesting activities.

• To determine the costs associated with alternative silvicultural systems, including values foregone when marketable wood is not harvested.

• To develop methodology for drastically improved and expanded monitoring systems (including habitat classification) necessary to assess performance of new silvicultural systems. This need includes the identification of suitable variables and the development of sampling and analytic procedures for assessing whether systems are achieving predicted management goals.

• To investigate impacts on future yields from alternative silvicultural systems.

Intensive Management for Wood Production

• To define sites that are especially well suited for efficient wood production through the use of habitat classification and other analytic tools.

• To investigate modifications in forestry practices that will enhance wildlife, water, scenery, and recreational values, and still allow efficient wood production.

• To develop technologies for harvesting, site preparation, regeneration, early plantation management, and protection of the forest from fire, insects, and disease that minimize the use of potentially dangerous or otherwise unacceptable forestry practices.

• To develop techniques of efficient and safe harvest of timber under complex silvicultural systems involving multiple entries, multilayer stands, partial cuts, and the leaving of woody debris on site.

• To determine the silvicultural regimes needed to grow wood of specified quality.

• To determine how genetic improvement of trees will modify silvicultural treatments and future growth and yields.

HUMAN-FOREST INTERACTIONS

Sociology and Forestry

• To undertake a systematic study of "human capital" for the twenty-first century with special attention to encouraging women and members of minority groups to conduct research and to manage forest systems.

• To establish the human and biological conditions required for urban forests to thrive.

• To assess the impact of urbanization at the forest edge.

• To examine the forest industry and dependent communities in transition, since the industry and its workers will continue to change.

• To broaden forest sociology research. Sociological research has been more heavily oriented toward management problems than toward understanding what motivates people to participate in recreational activities, patterns and cycles of use, and recreational trends.

• To expand research that deals with forest regions and forest communities. Demographic change, problems of fire, and residential locations on the fringes of USDA Forest Service areas merit research attention.

• To facilitate studies of private and public forest organizations undergoing change to better understand forest management practices, policy formation, and technology development.

• To develop educational programs for extension specialists on the utilization of sociological knowledge within extension programming; at the same time to encourage forest social scientists to seek careers in forestry extension.

WOOD AS A RAW MATERIAL

The Need for a Major and Sustained Commitment to Forest Products Research in the United States

• To understand certain tree species groups since U.S. industries will either use them or meet them in the marketplace. Worldwide, the man-made forests are increasingly growing certain staple forest crops, for example, the pines, eucalyptuses, teaks, and legumes.

- To develop essential technologies for adapting to changes in raw material. Roundwood from U.S. forests is increasingly of smaller size and higher in juvenile wood content.
- To utilize engineered raw material to improve future manufacturing efficiency. For example, the pulp and paper sector of the forest products industry has evolved from a "user of residuals" decades ago to a dominating role today.
- To gather knowledge of the structure and properties—physical and chemical—of wood, which is an essential base for improved product design, manufacture, and innovative applications.
- To employ biotechnology in developing new methods for converting existing low-value or residue products from existing wood-processing methods into products of high value. Of particular utility may be microorganisms that can convert low-value products into usable and economically feasible products; genetic engineering may improve the efficiency of these organisms. Microorganisms may also provide an improved means of cleaning up waste products of the industry.
- To improve understanding of the fundamental mechanisms of decay, formulate new approaches to decay control, and develop multifunctional chemical treatments that simultaneously contribute durability, weathering stability, decay resistance, and fire protection to wood products.
- To pursue manufacturing and process optimization research to increase product yield and manufacturing efficiency through computer-based technology, robotics, defect scanners, mechanized grading devices, and automated drying procedures.
- To develop and improve new pulping processes to mitigate environmental problems from pollutants such as dioxins and chloroform. Necessary, too, is the development of nonchlorine bleaching processes, high-value papers using polymeric additives, newsprint from hardwoods and waste paper, and methods to improve the performance of high-yield pulp.
- To improve methods to control costs and production so that labor and equipment can be used with maximum efficiency and cost effectiveness.
- To develop new and improved methods of quality control, which are essential if U.S. products are to be competitive in the marketplace.
- To increase the use of wood in nonresidential construction through new engineering design aides, streamlined engineering calculations, and simplified building code requirements.
- To develop new structural wooden assembly products and associated novel connector systems.
- To improve the recycling of discarded wood products, particularly paper, and to include solutions to subcomponent problems of paper recycling, such as de-inking.

- To enhance market analysis for wood and wood fiber products in order to create new and expanded opportunities in both domestic and international markets.

Timber Harvesting Research

- To understand relationships among engineering design, silviculture, environment, logger safety, and operating characteristics of the individual machine and the raw material production system (full rotation).
- To invent new systems, machines, and technologies to harvest timber under alternative silvicultural systems, safely, at acceptable costs, and with minimal environmental damage.
- To design harvesting systems that are integrated with silvicultural systems and that enhance all outputs from the forest.
- To develop measurement, sensor, and data-handling technology that will allow data to be more effectively used in engineering design and forestry operations.
- To develop ways of providing information feedback to managers of operations and operators of machines so that environmental, production, and safety goals can be met.
- To develop logger training curricula and materials, to test systems used to teach loggers, and to develop equipment that will be safer for loggers to use.

INTERNATIONAL TRADE, COMPETITION, AND COOPERATION

Information, Supply, and Demand

- To develop economic models that incorporate supply and demand information concerning natural resources; these models should project the resource consequences of national and international market strategies and policies.
- To collect, analyze, monitor, and disseminate data on domestic and worldwide natural resource inventories and international trade.
- To develop economic models that explore how international trade and debt policies can foster equity among nations in both benefits and costs of environmentally sound management of natural resources coupled with sustainable economic growth.
- To determine the true value of natural resources, focusing on both commodity and noncommodity outputs of the forest environment.
- To develop accounting systems for natural resources that accurately reflect the value of forest resources and the products thereof.

• To relate forest resources and their management to global warming. This research should investigate both effect of deforestation on global warming and the effects of global warming on forest resources.

• To identify nations or forest resource conditions that promise comparative economic advantage for expanded trade.

• To explore and analyze U.S. and international natural resource policies that will encourage sound land use, integrating forest management with watershed management to reduce erosion, lessen stream flow alterations, and reduce water pollution.

• To establish regional International Trade and Development Centers (ITDCs) in forest products to overcome the lack of information and analysis that hinders promotion of exports.

• To address the issue of diminishing fuel wood supplies for growing segments of the world's population.

Biographical Information on Committee Members

JOHN C. GORDON (Chair) is dean and professor of the School of Forestry and Environmental Studies at Yale University. His research includes photosynthesis and translocation in trees; enzymes in woody plants, and biological nitrogen fixation. Dr. Gordon received his undergraduate and doctoral degrees from Iowa State University and is a member of the American Association for the Advancement of Science (AAAS), Phi Kappa Phi, Sigma Xi, and the Society of American Foresters.

WILLIAM A. ATKINSON serves as professor and head of the Department of Forest Engineering at Oregon State University and as director of the OSU Research Forest. Dr. Atkinson has also held teaching positions at the University of California, Berkeley, and the University of Washington in addition to management and research positions in the forest products industry. He received his bachelor's, master's, and doctoral degrees from the University of California, Berkeley.

ELLIS B. COWLING is associate dean for research in the College of Forest Resources and University Distinguished Professor of Natural Resources at North Carolina State University. A member of the National Academy of Sciences and the NRC Board on Agriculture, Dr. Cowling does research on changes in the chemical climate of the earth and their impact on terrestrial and aquatic ecosystems, forest and wood products pathology, and physiology of trees and tree diseases. He received his B.S. and M.S. degrees from the State University of New York, Syracuse, and his Ph.D. from the University of Wisconsin and the University of Uppsala in Sweden.

MARY L. DURYEA received her B.S. and M.S. from the University of California, Berkeley, and her Ph.D. in tree physiology from Oregon

State University; she is currently an assistant professor in the Department of Forestry at the University of Florida. Her research interests include seed biology, nursery practices, and nutrition. Dr. Duryea is a member of Sigma Xi, Phi Beta Kappa, the Society of American Foresters, and is editor-in-chief of *New Forests*.

GEORGE F. DUTROW is dean and professor of forest economics in the Duke University School of Forestry and Environmental Studies. Dr. Dutrow previously taught at the University of Georgia and Our Lady of Holy Cross College and has been employed by the USDA Forest Service. He was awarded his bachelor's, master's, and doctoral degrees by Duke University and is a member of the Society of American Foresters.

DONALD R. FIELD is associate dean of the College of Agricultural and Life Sciences and director of the School of Natural Resources at the University of Wisconsin, Madison, where he received his bachelor's and master's degrees in rural sociology. He received his Ph.D. from Pennsylvania State University. In addition to experience with the National Park Service, Dr. Field has held teaching positions with South Dakota State University, the University of Washington, and Oregon State University. His research includes studies of the social ecology of parks and the impacts of rural resource development activities on communities and their regions.

RICHARD F. FISHER serves as professor and head of the Department of Forest Resources at Utah State University. His research includes studies on soil-plant relationships, plant-plant interactions, soil chemistry and bio-chemistry, and nitrogen fixation. He is a fellow of the Soil Science Society of America and the Society of American Foresters and is co-editor-in-chief of *Forest Ecology and Management*. Dr. Fisher received his B.S. from the University of Illinois and his Ph.D. from Cornell University.

JERRY F. FRANKLIN is the Bloedel Professor of Ecosystems Analysis in the College of Forest Resources at the University of Washington. He is also a chief plant ecologist for the USDA Forest Service. A member of AAAS, the Society of American Foresters, and the Ecological Society of America, Dr. Franklin conducts research on ecosystem structure and function, forest community ecology and succession, effects of environmental change, and incorporation of biological diversity into forest management.

DAVID W. FRENCH is a professor in the departments of Plant Pathology and Forest Resources at the University of Minnesota, St. Paul. Dr. French received his undergraduate, graduate, and doctoral degrees from the University of Minnesota, and his research focuses on products pathology, mycology, and forest pathology.

WILLIAM T. GLADSTONE received a B.S. from Syracuse University, an M.F. from Yale University, and a Ph.D. from North Carolina State University. His most recent position was manager of the Southern Forestry Research Department for Weyerhaeuser Company. His research includes variability and heritability of wood properties and relationships between wood fiber properties and products manufactured from wood. Dr. Gladstone is a member of the Technical Association of Pulp and Paper Industry, the Forest Products Research Society, and the Society of American Foresters.

LAWRENCE D. HARRIS serves as professor of wildlife ecology in the School of Forest Resources and Conservation at the University of Florida. In past positions, Dr. Harris has served as a wildlife management officer with the Tanzania Game Division and a wildlife biologist in the United States. His primary research is on development of renewable resource management strategies. A member of the Society for Conservation Biology, AAAS, Sigma Xi, and the Wildlife Society, he earned his B.Sc. from Iowa State University and his M.Sc. and Ph.D. from Michigan State University.

LOIS K. MILLER is a professor in the departments of Entomology and Genetics at the University of Georgia. A member of the American Society for Microbiology and the Society of Invertebrate Pathology, Dr. Miller does research in such areas as nucleic acid biochemistry, molecular biology, recombinant DNA technology, and biological insect pest control. She received her B.S. from Uppsala College and a Ph.D. in biochemistry from the University of Wisconsin, Madison.

JAMES R. SEDELL is a research aquatic ecologist at the Pacific Northwest Forestry and Range Experiment Station for the USDA Forest Service. Dr. Sedell has held positions with Oregon State University and Weyerhaeuser Company. He received his B.A. from Willamette University and his Ph.D. in aquatic biology from the University of Pittsburgh. He is a member of the American Fisheries Society, the Ecological Society of America, the North American Benthological Society, and the Society of American Foresters.

RONALD R. SEDEROFF serves as a professor of forestry at North Carolina State University. His research is on the molecular genetics of conifers. Dr. Sederoff earlier served as a senior scientist with the USDA Forest Service and has also held a variety of teaching positions. He received his bachelor's, master's, and doctoral degrees in zoology from the University of California, Los Angeles.

DAVID B. THORUD is the dean of the College of Forest Resources at the University of Washington. Dr. Thorud has also been employed by the USDA Forest Service and has held a number of teaching positions. He has participated in international delegations on such issues as watershed management training and research, soil and water conservation, and rural development. Dr. Thorud is a graduate of the University of Minnesota, where he received his B.S. in forestry and his M.S. and Ph.D. in forest hydrology.

Membership Rosters

BOARD ON BIOLOGY

FRANCISCO J. AYALA (*Chairman*), University of California, Irvine
NINA V. FEDOROFF, Carnegie Institution of Washington, Baltimore, Maryland
TIMOTHY H. GOLDSMITH, Yale University, New Haven, Connecticut
RALPH W. F. HARDY, Boyce Thompson Institute for Plant Research, Ithaca, New York
ERNEST G. JAWORSKI, Monsanto Company, St. Louis, Missouri
HAROLD A. MOONEY, Stanford University, Stanford, California
HAROLD J. MOROWITZ, George Mason University, Fairfax, Virginia
MARY-LOU PARDUE, Massachusetts Institute of Technology, Cambridge
DAVID D. SABATINI, New York University, New York
MICHAEL E. SOULÉ, University of California, Santa Cruz
MALCOLM S. STEINBERG, Princeton University, Princeton, New Jersey
DAVID B. WAKE, University of California, Berkeley
BRUCE M. ALBERTS (*ex-officio*), University of California, San Francisco

NRC Staff

OSKAR R. ZABORSKY, *Director*

COMMISSION ON LIFE SCIENCES

BRUCE M. ALBERTS (*Chairman*), University of California, San Francisco

BRUCE N. AMES, University of Calfornia, Berkeley

FRANCISCO J. AYALA, University of California, Irvine

J. MICHAEL BISHOP, University of California Medical Center, San Francisco

FREEMAN J. DYSON, Institute for Advanced Study, Princeton, New Jersey

NINA V. FEDOROFF, Carnegie Institution of Washington, Baltimore, Maryland

RALPH W. F. HARDY, Boyce Thompson Institute for Plant Research, Ithaca, New York

LEROY E. HOOD, California Insittute of Technology, Pasadena

DONALD F. HORNIG, Harvard School of Public Health, Boston, Massachusetts

ERNEST G. JAWORSKI, Monsanto Company, St. Louis, Missouri

MARIAN E. KOSHLAND, University of California, Berkeley

HAROLD A. MOONEY, Stanford University, Stanford, California

STEVEN P. PAKES, University of Texas Southwestern Medical School, Dallas

JOSEPH E. RALL, National Institutes of Health, Bethesda, Maryland

RICHARD D. REMINGTON, University of Iowa, Iowa City

PAUL G. RISSER, University of New Mexico, Albuquerque

RICHARD B. SETLOW, Brookhaven National Laboratory, Upton, New York

TORSTEN N. WIESEL, Rockefeller University, New York, New York

NRC Staff

JOHN E. BURRIS, *Executive Director*

BOARD ON AGRICULTURE

THEODORE L. HULLAR (*Chairman*), University of California, Davis
C. EUGENE ALLEN, University of Minnesota, St. Paul
EDWIN H. CLARK, State of Delaware Department of Natural Resources and Environmental Control, Dover
R. JAMES COOK, USDA Agricultural Research Service, Washington State University, Pullman
ELLIS B. COWLING, North Carolina State University, Raleigh
JOSEPH P. FONTENOT, Virginia Polytechnic Institute and State University, Blacksburg
ROBERT M. GOODMAN, Calgene, Inc., Davis, California
TIMOTHY M. HAMMONDS, Food Marketing Institute, Washington, D.C.
PAUL W. JOHNSON, State House of Representatives, Des Moines, Iowa
JOHN W. MELLOR, International Food Policy Research Institute, Washington, D.C.
CHARLES C. MUSCOPLAT, Molecular Genetics, Inc., Minnetonka, Minnesota
KARL H. NORRIS, Beltsville, Maryland
CHAMP B. TANNER, University of Wisconsin, Madison
ROBERT L. THOMPSON, Purdue University, West Lafayette, Indiana
JAN VAN SCHILFGAARDE, USDA Agricultural Research Service, Fort Collins, Colorado
ANNE M. K. VIDAVER, University of Nebraska, Lincoln
CONRAD J. WEISER, Oregon State University, Corvallis

NRC Staff

CHARLES M. BENBROOK, *Executive Director*
JAMES E. TAVARES, *Associate Executive Director*

Index

Date Due

APR 8 2004			